# Urban Water Systems
# & Floods III

**WIT**PRESS

WIT Press publishes leading books in Science and Technology.
Visit our website for the current list of titles.
www.witpress.com

**WIT**eLibrary

Home of the Transactions of the Wessex Institute.
Papers published in this volume are archived in the WIT eLibrary in volume 194 of WIT Transactions on the Built Environment (ISSN 1743-3509). The WIT eLibrary provides the international scientific community with immediate and permanent access to individual papers presented at WIT conferences.
http://library.witpress.com.

SEVENTH INTERNATIONAL CONFERENCE ON
FLOOD AND URBAN WATER MANAGEMENT

# FRIAR 2020

## CONFERENCE CHAIRMEN

**S. Mambretti**
*Polytechnic of Milan, Italy*
*Member of WIT Board of Directors*

**D. Proverbs**
*Birmingham City University UK*

## INTERNATIONAL SCIENTIFIC ADVISORY COMMITTEE

## ORGANISED BY

*Wessex Institute, UK*
*Birmingham City University, UK*
*Polytechnic of Milan, Italy*

## SPONSORED BY

*WIT Transactions on the Built Environment*
*International Journal of Environmental Impacts*

# WIT Transactions

Wessex Institute
Ashurst Lodge, Ashurst
Southampton SO40 7AA, UK

## Senior Editors

## Editorial Board

K. **Dorow** Pacific Northwest National Laboratory, USA

W. **Dover** University College London, UK

C. **Dowlen** South Bank University, UK

J. P. **du Plessis** University of Stellenbosch, South Africa

R. **Duffell** University of Hertfordshire, UK

A. **Ebel** University of Cologne, Germany

V. **Echarri** University of Alicante, Spain

K. M. **Elawadly** Alexandria University, Egypt

D. **Elms** University of Canterbury, New Zealand

M. E. M **El-Sayed** Kettering University, USA

D. M. **Elsom** Oxford Brookes University, UK

F. **Erdogan** Lehigh University, USA

J. W. **Everett** Rowan University, USA

M. **Faghri** University of Rhode Island, USA

R. A. **Falconer** Cardiff University, UK

M. N. **Fardis** University of Patras, Greece

A. **Fayvisovich** Admiral Ushakov Maritime State University, Russia

H. J. S. **Fernando** Arizona State University, USA

W. F. **Florez-Escobar** Universidad Pontifica Bolivariana, South America

E. M. M. **Fonseca** Instituto Politécnico do Porto, Instituto Superior de Engenharia do Porto, Portugal

D. M. **Fraser** University of Cape Town, South Africa

G. **Gambolati** Universita di Padova, Italy

C. J. **Gantes** National Technical University of Athens, Greece

L. **Gaul** Universitat Stuttgart, Germany

N. **Georgantzis** Universitat Jaume I, Spain

L. M. C. **Godinho** University of Coimbra, Portugal

F. **Gomez** Universidad Politecnica de Valencia, Spain

A. **Gonzales Aviles** University of Alicante, Spain

D. **Goulias** University of Maryland, USA

K. G. **Goulias** Pennsylvania State University, USA

W. E. **Grant** Texas A & M University, USA

S. **Grilli** University of Rhode Island, USA

R. H. J. **Grimshaw** Loughborough University, UK

D. **Gross** Technische Hochschule Darmstadt, Germany

R. **Grundmann** Technische Universitat Dresden, Germany

O. T. **Gudmestad** University of Stavanger, Norway

R. C. **Gupta** National University of Singapore, Singapore

J. M. **Hale** University of Newcastle, UK

K. **Hameyer** Katholieke Universiteit Leuven, Belgium

C. **Hanke** Danish Technical University, Denmark

Y. **Hayashi** Nagoya University, Japan

L. **Haydock** Newage International Limited, UK

A. H. **Hendrickx** Free University of Brussels, Belgium

C. **Herman** John Hopkins University, USA

I. **Hideaki** Nagoya University, Japan

W. F. **Huebner** Southwest Research Institute, USA

M. Y. **Hussaini** Florida State University, USA

W. **Hutchinson** Edith Cowan University, Australia

T. H. **Hyde** University of Nottingham, UK

M. **Iguchi** Science University of Tokyo, Japan

L. **Int Panis** VITO Expertisecentrum IMS, Belgium

N. **Ishikawa** National Defence Academy, Japan

H. **Itoh** University of Nagoya, Japan

W. **Jager** Technical University of Dresden, Germany

Y. **Jaluria** Rutgers University, USA

D. R. H. **Jones** University of Cambridge, UK

N. **Jones** University of Liverpool, UK

D. **Kaliampakos** National Technical University of Athens, Greece

D. L. **Karabalis** University of Patras, Greece

A. **Karageorghis** University of Cyprus

T. **Katayama** Doshisha University, Japan

K. L. **Katsifarakis** Aristotle University of Thessaloniki, Greece

E. **Kausel** Massachusetts Institute of Technology, USA

H. **Kawashima** The University of Tokyo, Japan

B. A. **Kazimee** Washington State University, USA

F. **Khoshnaw** Koya University, Iraq

S. **Kim** University of Wisconsin-Madison, USA

D. **Kirkland** Nicholas Grimshaw & Partners Ltd, UK

E. **Kita** Nagoya University, Japan

A. S. **Kobayashi** University of Washington, USA

D. **Koga** Saga University, Japan

S. **Kotake** University of Tokyo, Japan

# Urban Water Systems & Floods III

### Editors

**S. Mambretti**
*Polytechnic of Milan, Italy*
*Member of WIT Board of Directors*

**D. Proverbs**
*Birmingham City University UK*

**WIT**PRESS Southampton, Boston

**Editors:**

**S. Mambretti**
*Polytechnic of Milan, Italy*
*Member of WIT Board of Directors*

**D. Proverbs**
*Birmingham City University UK*


Published by

**WIT Press**
Ashurst Lodge, Ashurst, Southampton, SO40 7AA, UK
Tel: 44 (0) 238 029 3223; Fax: 44 (0) 238 029 2853
E-Mail: witpress@witpress.com
http://www.witpress.com

For USA, Canada and Mexico

**Computational Mechanics International Inc**
25 Bridge Street, Billerica, MA 01821, USA
Tel: 978 667 5841; Fax: 978 667 7582
E-Mail: infousa@witpress.com
http://www.witpress.com

British Library Cataloguing-in-Publication Data

A Catalogue record for this book is available
from the British Library

ISBN: 978-1-78466-379-7
eISBN: 978-1-78466-380-3
ISSN: 1746-4498 (print)
ISSN: 1743-3509 (on-line)

*The texts of the papers in this volume were set individually by the authors or under their supervision. Only minor corrections to the text may have been carried out by the publisher.*

# Preface

This volume contains a selection of papers presented at the 7th International Conference on Flood and Urban Water Management, which, scheduled in Valencia, Spain, was held on line due to the Coronavirus pandemic. The Conference has been organised by the Wessex Institute, in collaboration with the Politecnico di Milano in Italy and Birmingham City University in the UK.

FRIAR 2020 is the seventh conference of this successful series. The conference started at the Institution of Civil Engineers in London 2008 and was reconvened in Milan in 2010, Dubrovnik in 2012, Poznan in 2014, Venice in 2016 and A Coruña 2018. Since 2012 a parallel seminar on the Design, Construction, Maintenance, Monitoring and Control of Urban Water has taken place which is now an integral part of the conference.

Flooding is a global phenomenon that claims numerous lives worldwide each year. The primary effects of flooding include loss of life and damage to buildings and other structures, including bridges, sewerage systems, roadways and canals. Floods also frequently damage power transmission and sometimes power generation, which effects include loss of drinking water treatment and water supply, which may result in loss of drinking water or severe water contamination. It may also cause the loss of sewage disposal facilities. Lack of clean water combined with human sewage in the flood waters raises the risk of waterborne diseases. Damage to roads and transport infrastructure may make it difficult to mobilize aid to those affected or to provide emergency health treatment.

The increased frequency of flooding in the last few years, coupled with climate change predictions and urban development, suggest that these impacts are set to worsen in the future. How we respond, and importantly adapt, to these challenges is key to developing our long-term resilience at the property, community and city scale.

Apart from the physical damage to buildings, contents and loss of life, which are the most obvious, impacts of floods upon households, other more indirect losses are often overlooked. These indirect and intangible impacts are generally associated with disruption to normal life as well as longer-term health issues including community displacements and stress-related illnesses. Flooding represents a major barrier to the alleviation of poverty in many parts of the developing world, where vulnerable communities are often exposed to sudden and life-threatening events.

Planning for flood safety involves many aspects of analysis and engineering, including observation of previous and present flood heights and inundated areas, statistical, hydrologic, and hydraulic model analyses, mapping inundated areas and flood heights for future flood scenarios, long-term land use planning and regulation, engineering design and construction of structures to control or withstand flooding, intermediate-term monitoring, forecasting, and emergency-response planning, and short-term monitoring, warning and response operations.

Defences such as detention basins, levees, bunds, reservoirs, and weirs are used to prevent waterways from overflowing their banks. When these defences fail, emergency measures have to be designed.

As our cities continue to expand, their urban infrastructures need to be re-evaluated and adapted to new requirements related to the increase in population and the growing areas under urbanization. This conference also considers these problems and deals with two main urban water topics: water supply systems and urban drainage.

Topics such as contamination and pollution discharges in urban water bodies, as well as the monitoring of water recycling systems are currently receiving a great deal of attention from researchers and professional engineers working in the water industry. Water distribution networks often suffer substantial losses which represent wastage of energy and treatment. Effective, efficient and energy saving management is necessary in order to optimize their performance. Sewer systems are under constant pressure due to growing urbanization and climate change, and the environmental impact caused by urban drainage overflows is related to both water quantity and water quality.

FRIAR seeks to develop an improved understanding of emerging flood risk management and urban water management challenges, drawing on the expertise of numerous disciplines and considering a range of responses. The conference provides a rich forum for the development of innovative solutions that can help bring about multiple benefits toward achieving integrated flood risk and urban water management strategies and policy.

The meeting attracted researchers, academics and practitioners actively involved in improving our understanding of urban water systems and flood events. It brings together social scientists, surveyors, engineers, scientists, and other professionals from many countries involved in research and development activities in a wide range of technical and management topics related to urban water and flooding and its impacts on communities, property and people.

The papers included in this volume make an improved understanding of emerging flood risk management and urban water management challenges, drawing on the expertise of numerous disciplines and considering a range of responses. The conference provided a rich forum for the development of innovative solutions that can help bring about multiple benefits toward achieving integrated flood risk and urban water management strategies and policy.

These papers, like others presented at Wessex Institute conferences, are referenced by CrossRef and appear regularly in suitable reviews, publications and databases, including referencing and abstracting services. They are also archived online in the WIT eLibrary (http://www.witpress.com/elibrary) where they are permanently available in Open Access format to the international scientific community.

The Editors would like to thank the authors for their contributions, as well as the member of the International Scientific Advisory Community of the Conference for their invaluable help in reviewing the papers.


The Editors, 2020

# Contents

# SECTION 1
# FLOOD RISK
# MANAGEMENT

# FLOOD ENVIRONMENTAL IMPACT RISK ANALYSIS

MARTINA ZELEŇÁKOVÁ[1], MÁRIA ŠUGAREKOVÁ[1] & PETER MÉSÁROŠ[2]
[1]Department of Environmental Engineering, Technical University of Košice, Slovakia
[2]Department of Construction Technology and Management, Technical University of Košice, Slovakia

## ABSTRACT

The aim of this paper is to determine the degree of threat in the floodplain of the Hornád river basin and to propose measures that can be used in practice at the time of the flood. The methodology is based on the principle of FEIRA (Flood Environmental Impact Risk Analysis), where the probability and consequence of the negative impact of floods on environmental components is determined on the basis of the analysis of selected stressors. From these indicators, the degree of risk in the Hornád river basin in the event of a flood is subsequently determined. Since there are several industrial sites in the solved area, which can cause extensive pollution of watercourses in case of floods, the work also presents the calculation of the threat according to the point evaluation of pollution sources. The combination of the FEIRA process and the proposed methodology for the assessment of flood environmental damage determined the overall risk of environmental damage due to floods in the Hornád catchment area. This paper is a proposal for measures to protect against floods in the area in case of floods.
*Keywords: flood, environmental impact analysis, risk analysis, Hornád river basin.*

## 1 INTRODUCTION

A wide range of models and case studies are used to assess flood risks. However, internationally, these models have significant economic differences. The authors Jongman et al. [1] call for the development of a consistent European framework that will apply procedures from existing models. The assessment of the economic and social impacts of floods can also be carried out using models made from topographic data of the area [2]. At the beginning of autumn 2003, a flood risk assessment study was carried out in the German Research System for Natural Disasters [3]. Flood risk is on the rise [4]–[6]. A study by Tincu et al. [7] estimates the direct damage caused by the occurrence of three floods at different times, which speaks of emphasizing the need to improve spatial plans. A study by Mishra and Sinha [8] proves that floods are one of the most devastating natural disasters, causing enormous damage to property and, in some cases, loss of life. Climate change has an impact not only on individual components of the environment, but also on the social sphere [9].

Slovakia belongs to the countries that are increasingly being affected by floods [10], [11]. In this work, the main goal is to identify and assess the impacts and impact of floods on environmental components. The impact and assessment is focused on the territory of the Hornád river basin located in the territory of the Slovak Republic. The case studies describe the methods for dealing with flood impact assessments in some countries, as well as the methods used, resp. proposed solution methodology. The following section is devoted to the description of the solved area: the partial catchment area of Hornád, to which the practical part of this work applies.

## 2 STUDY AREA

The Hornád river basin (Fig. 1) covers the territory of the Slovak and Hungarian Republics with a total area of 4,414 km$^2$ and a length of 193 km in Slovakia and 93 km in Hungary.

The area of the Hornád sub-basin is characterized by the occurrence of impermeable and poorly permeable rocks, which have a moderate to low permeability. Rocks with good to high flow can be found in the Košice Basin and in the areas of the Slovak Karst and the

WIT Transactions on The Built Environment, Vol 194, © 2020 WIT Press
www.witpress.com, ISSN 1743-3509 (on-line)
doi:10.2495/FRIAR200011

Spiš-Gemer region. The rocks in the area of the Slovak Ore Mountains and Branisko are poorly permeable with a predominant fissure permeability. Atmospheric precipitation is the main source of groundwater in this area [12].

Part of the Hornád basin is located in the province of the Western Carpathians, a sub-province of the Inner Western Carpathians, and a smaller part of it belongs to the sub-province of the Outer Western Carpathians.

The sub-province of the Inner Western Carpathians includes the area:

- Slovak Ore Mountains,
- Fatra-Tatra region,
- Lučensko-Košice reduction,
- Matransko-Slanská area.

The areas belonging to the sub-province of the Outer Western Carpathians are the Eastern Beskydy and the Podhôlno-Magurská area.

The highest altitude of 1,401 m asl. occupies the smallest area of the solved area, the largest area lies at an altitude of 300–500 m asl. [12].

Due to the geographical location of the Hornád sub-basin, there are up to 3 climatic areas. The warm to slightly dry climatic part, characterized by a cold winter, includes the southern to south-eastern area bordering the Sabinov area. The middle part of the basin is characterized by an average total precipitation from 700–900 mm. The climatic conditions in this district are slightly warm, slightly humid to humid. The area bordering the Volovské vrchy is characterized by a slightly cold climate. The air temperature here ranges from 4–5°C. The annual total precipitation also exceeds 900 mm [12].

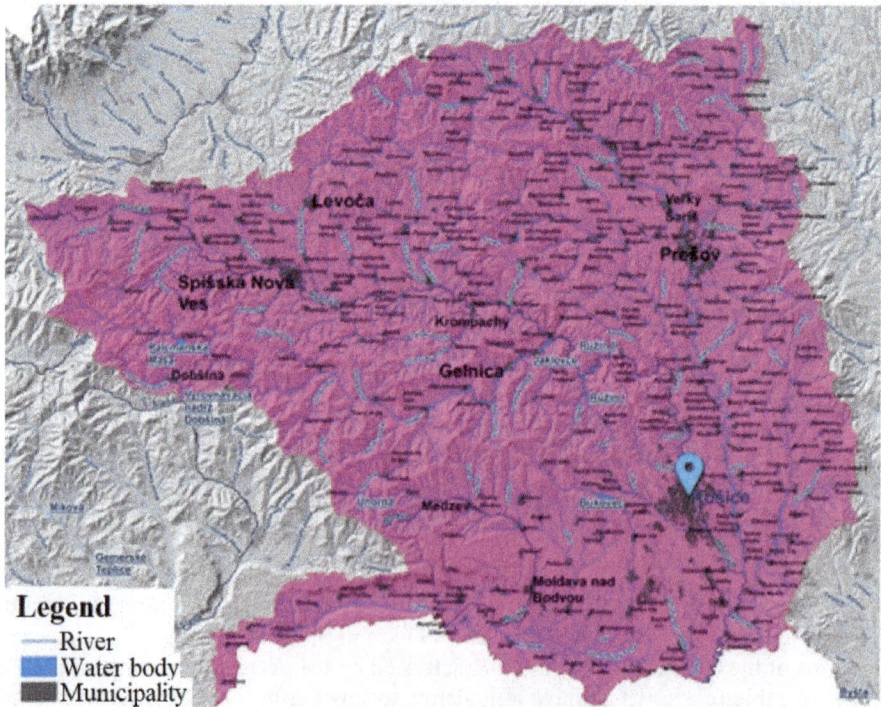

Figure 1:  Study area: Hornád river basin.

The Hornád river basin covers an area of two states. In the territory of the Slovak Republic, together with its tributaries, it occupies 81% of its total area. The most important right-hand tributary is the river Hnilec, which flows near the village of Margecany. Hnilec springs on the slope of Kráľová hoľa at an altitude of approximately 1740 m asl. It flows into the Palcmanská Maša reservoir, which is also the largest reservoir in the Slovak Paradise National Park. The flow of Hnilec continues from the dam in an easterly direction to the Ružín Reservoir.

The most significant left tributary is created by the river Torysa near the village of Nižná Myšľa. It springs in Levočské vrchy, northwest of the village Torysky at an altitude of approximately 1215 m asl. Torysa flows south, flowing through the city of Prešov [12].

The Hornád river basin has a wide representation of different soil groups of geographically related soils. In the forest parts of the basin, there are mostly acidic varieties of cambium soils. These soils are very skeletal. Soils that occupy a large area in the Hornád basin are characterized by an acidic to strongly acidic soil reaction. These are soils with an insatiable sorption complex. In the Torysa basin, pseudogleies in particular are widespread. In this area, pseudogleys are arranged alternately – sorption or acid glues [12].

In the western part of the Hornád catchment area in the vicinity of Spiš, there is the Slovak Paradise mountain range, which is classified as a national park. The territory is located in the cadastral territory of the village Spišské Tomášovce. This location is rich in remarkable natural and historical monuments. The area of the Slovak Paradise National Park is 328 km$^2$, as a protected landscape area was determined in 1964. There are 11 national nature reserves and 8 nature reserves on the territory of the Slovak Paradise [12].

The following part of the work is devoted to the methodology of assessment and analysis of individual factors affecting selected components of the environment. The final part summarizes the final values of the assessment and the overall assessment of the impact of floods on the area, as well as the assessment of flood risk.

There are several industrial sites in the Hornád sub-basin, which may pose a risk of environmental pollution in the event of floods. Summary Table 1 lists these sources of pollution. In the event of a flood at Q100, several industrial sites pose a direct threat to humans but also to the environment. Namely, there are:

- Area of VSE, Company of mechanical production Krompachy, group unclassified;

- Wastewater treatment plants (WWTP): WWTP Harichovce, WWTP Vajkovce, AGROKOV PLUS Košice, WWTP Rožkovany, WWTP Jakubova Voľa – group of wastewater treatment plants up to 2,000 equivalent populations;

- WWTP Spišské Vlachy, WWTP Flood yard Krompachy, WWTP IMUNA PHARM Šarišské Michaľany – group of wastewater treatment plants from 2,000 to 10,000 equivalent populations;

- WWTP Spišská Nová Ves – group of wastewater treatment plants from 10,000 to 100,000 equivalent populations;

- KOVOHUTY Krompachy – group probable environmental burden,

- SEZ Krompachy – Electrical Production Plant – group of remediated or reclaimed locality.

Each pollution source is assigned an appropriate number of points according to the categorization.

## 3  MATERIAL AND METHODS

Flood risk assessment according to the FEIRA process (Flood Environmental Impact Risk Analysis) consists of the following steps [13]–[15]. FEIRA is based on the methodology of procedures defined in ISO 31000 – Risk Management – Principles and Guidance. The FEIRA process begins with a description of the current state of the environment and the definition of sources of pollution in the area. The next step is the identification of stressors – sources of risk that pose a danger in the assessed area and represent a potential impact on environmental components. The consequence is determined by the significance of the stressor's action on the selected evaluated component. The sum of the products of these two indicators is the obtained value of the risk index [11].

The IR (risk index) is measured by the product of the probability and the consequence of individual stressors expressed by the following eqn:

$$IR = Pi * Ci. \tag{1}$$

The probability Pi expresses the value of each selected stressor effect, and the consequence Ci expresses the value of the stressor effect on the individual components.

The value of the total risk posed by flooding in the event of floods is determined by the following eqn:

$$R = SUM (Pi*Di) * Hi. \tag{2}$$

The hazard Hi presents the sources of pollution in the river basin [12], [13].

The goal of this contribution is effective flood risk assessment and management in the studied location which is based on current state of the environment and presence of sources of pollution in the study area. According to the proposed methodology the there are three steps:

- To calculate risk index which presents impact of stressors (floods) on components of the environment based on determined probability and determined consequence.
- To calculate risk based on hazards in the area; hazard presents point or diffuse sources of pollution in the area.
- The selection of effective measures for flood protection.

The result is the proposal of possible flood protection measures which will be effective from the viewpoint of protection (economic as well as environmental).

## 4  RESULTS AND DISCUSSION

Probability is an expression of the possibility of the occurrence of a certain phenomenon. The starting point of this methodology is the qualitative determination of probability from the lowest value = 0.25 to the highest value 1. Level, resp. the probability value represents a certain expectation – whether the phenomenon will happen or can happen. How the negative stressor affects selected components of the environment is expressed by the consequence. It is expressed qualitatively – as well as probability.

All assessed effects with values of causes Pi and their consequences Ci are summarized in Table 1.

According to the proposed category, the total risk level in the addressed area of the Hornád river basin is determined by a value equal to 5.5 – medium risk.

If a flood occurs in any area, it can also cause pollution of environment. There are several industrial sites in the Hornád sub-basin, which may pose a risk of environmental pollution in the event of floods. According to the categorization, each source of pollution is assigned the appropriate number of points – the Hi score (Table 2).

Table 1:   Summary of the probabilities and consequences of the impact of the flood on the environment.

| ID | Impact of stressors on components of the environment | Determination of probabilities | | Determination of consequences | |
|---|---|---|---|---|---|
| 1 | Impact of flooding on the population | P1 | Local potential for flooding (-) | C1 | Health consequences of flooding (point) |
| | | 0.5 | medium | 1 | ≥ 5 |
| 2 | Impact of flooding on water conditions | P2 | Number of announcements of highest level of flooding (per year) (-) | C2 | Capacity flow Qn (m3.s-1) |
| | | 1 | > 4 | 0.75 | ≥ Q50 |
| 3 | Impact of flooding on soil | P3 | The status of flood protection facilities (-) | C3 | Permeability of soil (-) |
| | | 0.5 | good | 0.5 | Less permeable |
| 4 | Impact of flooding on flora, fauna and their habitats | P4 | Local potential for flooding (-) | C4 | Vulnerability of fauna and flora and their habitats (-) |
| | | 0.5 | medium | 0.75 | medium |
| 5 | Impact of flooding on landscape – structure and land use, landscape character | P5 | Local potential for flooding (-) | C5 | Changes in the landscape (-) |
| | | 0.5 | medium | 0.75 | significant |
| 6 | Impact of flooding on protected areas and their buffer zones | P6 | Local potential for flooding (-) | C6 | Location of the proposed activity (-) |
| | | 0,5 | medium | 1 | within 3 and more protection areas |
| 7 | Impact of the flooding of the territorial system of ecological stability (TSES) | P7 | The status of flood protection facilities (-) | C7 | Impacts on TSES (point) |
| | | 0.5 | good | 0.5 | 6–10 |

Table 1: Continued.

| ID | Impact of stressors on components of the environment | | Determination of probabilities | | Determination of consequences |
|---|---|---|---|---|---|
| 8 | Impact of flooding on urban areas and land use | P8 | Local potential for flooding (-) | C8 | Local potential for flooding (-) |
| | | 0.5 | medium | 1 | ≥ 101 |
| 9 | Impact of flooding on cultural and historical heritage, intangible cultural values | P9 | Number of announcements of highest level of flooding (per year) (-) | C9 | Number of affected values in the area (-) |
| | | 1 | > 4 | 1 | ≥ 6 |
| 10 | Impact of flooding on archaeological and paleontological sites and important geological sites | P10 | Number of announcements of highest level of flooding (per year) (-) | C10 | Number of affected sites in the area (-) |
| | | 1 | > 4 | 1 | ≥ 3 |
| | $\sum_{i=1}^{n=10} IRi$ | | | | 5.5 |

The sum of the values of individual causes and consequences represents the value IR = 5.5, which is subsequently multiplied by the value of danger Hi = 19. The product of these two values is the resulting value representing the total risk R = 104.5. It presents very low risk for the Hornád river basin.

The goal of selecting effective flood protection measures in the studied territory include:

- the removing of soil deposits from the water channel and vegetation from the bank of the watercourse, thus securing the overflow capacity of the watercourse,

- for the unaltered sections of the watercourse to make modifications, e.g. to reinforce the slopes of the water channel,

- if necessary construction of a reservoir above the town which lowers the maximum overflow during increased water stages.

The construction of reservoir – dry basin above the municipality seems to be the most effective flood protection measure in the area.

Table 2: Pollution sources with assigned number of points.

| Category of source of pollution | Source of pollution | Criteria | Point evaluation $H_i$ |
|---|---|---|---|
| Industrial enterprises | Area of VSE – Enterprise of mechanical production Krompachy | unclassified | 5 |
| Wastewater treatment plants (WWTP) | WWTP Harichovce, WWTP Vajkovce, AGROKOV PLUS Košice, WWTP Rožkovany, WWTP Jakubova Voľa | < 2,000 population equivalent | 1 |
| | WWTP Spišské Vlachy, WWTP Povodňového dvora Krompachy, WWTP IMUNA PHARM Šarišské Michaľany | 2,000 – 10,000 population equivalent | 2 |
| | WWTP Spišská Nová Ves | 10,000 – 100,000 population equivalent | 3 |
| Agriculture | Crop production | 10–40% of flooded area | 1 |
| Environmental burden | KOVOHUTY Krompachy | Environmental burden is likely | 3 |
| | SEZ Krompachy – Enterprise of electric production | Land reclamation | 3 |
| Urban areas | Population without sewerage | 10–40% from all population in the study area | 1 |
| Sum Σ | | | 19 |

## 5 CONCLUSION

The first part of the practical solution of this paper was devoted to the analysis of individual stressors – floods, that have a negative impact on selected components of the environment. For each stressor impact, the probability and consequence of its effect was determined. The values were then multiplied by each other and the sum of the resulting values represents the resulting risk index. The risk index in the addressed area of the Hornád river basin is determined by a value equal to 5.5, which represents a medium level of risk level.

In the next part, the hazard was state according to the proposed methodology [14], [15] – FRIAR. In the solved area, sources of pollution were identified, to which a point value was assigned. The resulting value represented the sum of the partial results. In the Hornád sub-basin, a low threat rate applies with a hazard value is 19.

The final part was devoted to determining the overall risk. The resulting value of risk is the product of the resulting values of probability, consequence and threat. According to the FEIRA methodology, the first category of risk level is estimated to 104.5, which means a very low level of flood risk.

## ACKNOWLEDGEMENTS
This work has been supported by the Slovak Research and Development Agency by supporting the project SK-PT-18-0008 and project SL-PL-18-0033. This work was supported by projects of the Ministry of Education of the Slovak Republic VEGA 1/0308/20 Mitigation of hydrological hazards – floods and droughts – by exploring extreme hydroclimatic phenomena in river basins.

## REFERENCES
[1]   Jongman B. et al., Comparative flood damage model assessment: Towards a European approach. *Natural Hazards and Earth System Sciences*, **12**, pp. 3733–3752, 2012.
[2]   Hall, J.W. et al., A methodology for national-scale flood risk assessment. *Proceedings of the Institution of Civil Engineers – Water and Maritime Engineering*, **156**(3), pp. 235–247, 2003.
[3]   Apel, H., Thieken, A.H., Merz, B. & Blöschl, G., Flood risk assessment and associated uncertainty. *Natural Hazards and Earth System Science*, **4**(2), pp. 295–308, 2004.
[4]   Figueiredo, R., Romao, X. & Paupério, E., Flood risk assessment of cultural heritage at large spatial scales: Framework and application to mainland Portugal. *Journal of Cultural Heritage*, **41**, pp. 1–12, 2019.
[5]   Hanák, T. & Korytárová, J., Risk zoning in the context of insurance: Comparison of flood, snow load, windstorm and hailstorm. *Journal of Applied Engineering Science*. **12**, pp. 137–144, 2014.
[6]   Korytárová, J, Šlezingr, M. & Uhmannová, H., Determination of potential damage to representatives of real estate property in areas afflicted by flooding. *Journal of Hydrology and Hydromechanics*, **55**, pp. 282–228, 2007.
[7]   Tincu, R., Zezere, J.L., Craciun, I., Lazar, G. & Lazar I., Quantitative micro-scale flood risk assessment in a section of the Trotus River, Romania. *Land Use Policy*, **89**, pp. 1–13, 2019.
[8]   Mishra, K. & Sinha, R., Flood risk assessment in the Kosi megafan using multi-criteria decision analysis: A hydro-geomorphic approach. *Geomorphology*, **350**, pp. 1–19, 2020.
[9]   Metz, F., Angst, M. & Fischer, M., Policy integration: Do laws or actors integrate issues relevant to flood risk management in Switzerland? *Global Environmental Change*, **61**, pp. 1–12, 2020.
[10]  Zeleňáková, M., Preliminary flood risk assessment in the Hornád watershed. *WIT Transactions on Ecology and the Environment*, WIT Press: Southampton and Boston, pp. 15–24, 2009.
[11]  Zeleňáková, M., Flood risk assessment and management in Slovakia. *WIT Transactions on Ecology and the Environment*, WIT Press: Southampton and Boston, pp. 61–69, 2011.

[12]   Ministry of the Environment of the Slovak Republic: Preliminary flood risk assessment in the Slovak Republic. Bratislava: MŽP SR, 2018. http://www.minzp.sk/files/sekcia-vod/hodnotenie-rizika-2018/phpr_sr2018.pdf. Accessed on: 5 Nov. 2019.

[13]   Zeleňáková, M. et al., Mitigation of the Adverse Consequences of Floods for Human Life, Infrastructure, and the Environment. *Natural Hazards Review,* **18**, pp. 1–15, 2017.

[14]   Zeleňáková, M., Gaňová, L. & Purcz, P., Flood risk assessment as part of flood defence. *SGEM 2012: 12th International Multidisciplinary Scientific GeoConference,* Vol. 3, Albena, Bulgaria, STEF92 Technology Ltd., pp. 679–686, 2012.

[15]   Zeleňáková, M., Assessment of flood vulnerability in Bodva catchment using multicriteria analysis and geographical information systems. *WIT Transactions on Ecology and the Environment,* WIT Press: Southampton and Boston, pp. 51–59, 2015.

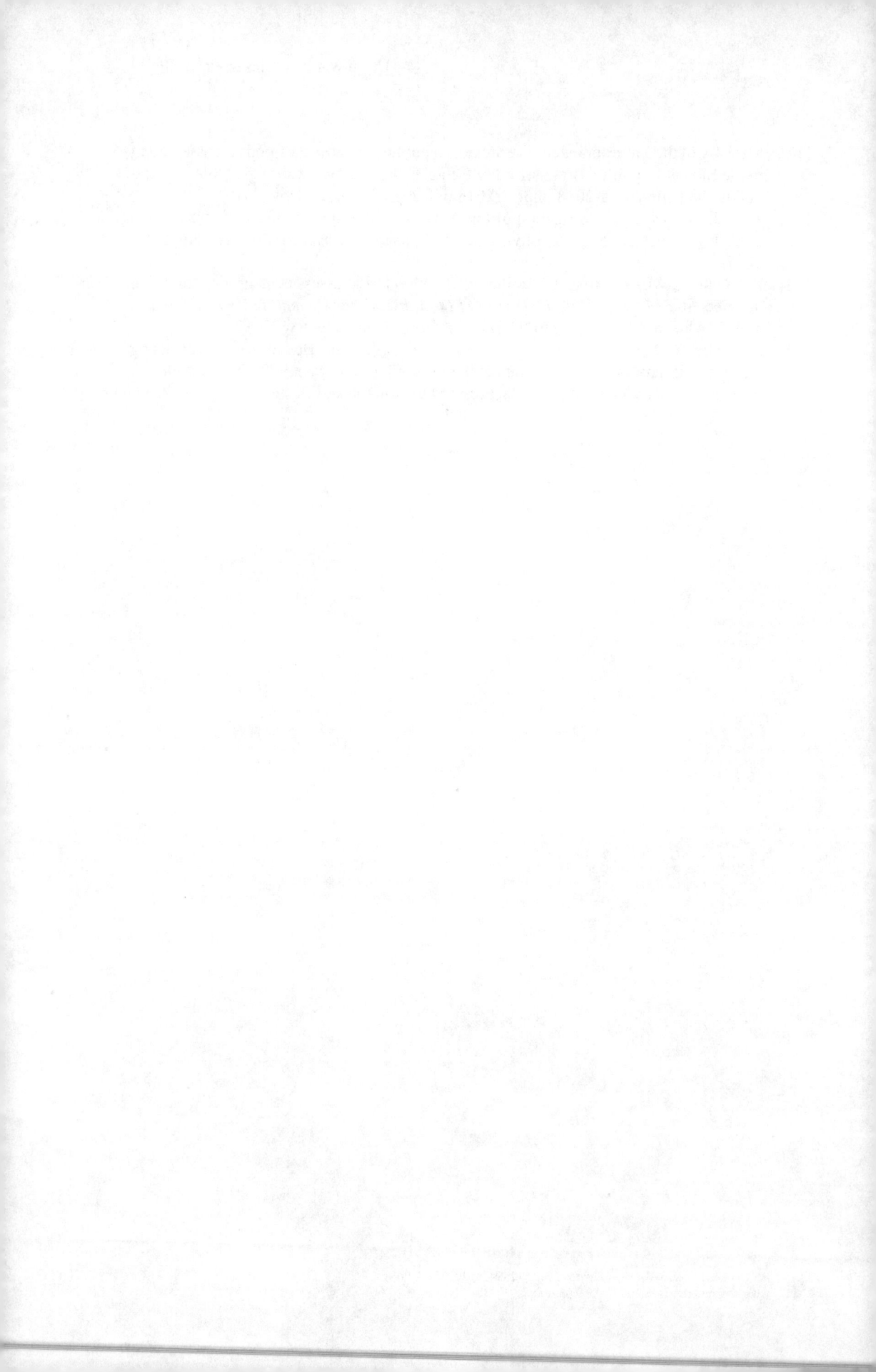

# IMPROVING THE FLOOD RESILIENCE OF COMMERCIAL BUILDINGS THROUGH PROPERTY FLOOD RESILIENCE MEASURES

HONG XIAO[1], DAVID PROVERBS[1], ROTIMI JOSEPH[2] & TAIWO ADEDEJI[1]
[1]Birmingham City University, UK
[2]University of Brighton, UK

## ABSTRACT

The impacts of flooding on businesses include financial damages, business interruption, breakdown of supplies and ultimately business failure. While some commercial properties have developed a level of resilience through taking steps to protect and adapt their premises, the majority are vulnerable to these impacts and lack any level of resilience to flooding. The concept of property flood resilience (PFR) involves the application of a range of measures that can be applied to a property to make people and their property less vulnerable to the impacts of flooding. While this approach has been the subject of much research, leading to an acceptance of this approach in UK flood risk policy and strategy, there has been a dearth of research on the use of PFR for commercial properties. The aim of this research is to explore the application of PFR to commercial properties, and to develop insights into their current usage as well as the potential application in the future. This research draws on a critical examination of the existing literature to reveal the full extent of the impacts of flooding on commercial properties. These impacts are classified as direct/indirect and tangible/intangible impacts, while mitigation measures are found to consist of a hierarchy based on avoidance, resistance, reliance and reparability. Further research is recommended on how to improve the flood resilience of commercial properties through property level measures.
*Keywords: commercial property, property flood resilience, flood impact, mitigation measures, adaptation measures.*

## 1 INTRODUCTION

Flood is one of the most wide-reaching and frequently occurring natural hazards in the world with noticeable impacts across cultures and geographies. On average, flood affects about 70 million people each year [1]. The impacts include physical damage to properties, critical infrastructures and assets. The losses caused as a result of business interruption and general disruption to communities is significant [2]. Also, the impacts on health are extensive and substantial, ranging from loss of lives and injuries sustained from the shock caused by the event, to the infectious diseases and mental health problems, including both acute and long-term. An analysis of global statistics conducted by Jonkman [3] revealed that floods had caused 175,000 fatalities and affected more than 2.2 billion people between 1975 and 2002.

Similarly, the impacts in the UK have been far-reaching with over 5.2 million properties (around one in seven homes and businesses) exposed to some form of flood risk [4]. Estimates suggest that over £220 billion worth of property is potentially at risk of flooding in England and Wales, from different sources of flooding such as coastal, rivers, surface water, groundwater and sewers flooding [5]. The coastal profile, areas within 10 km of the coast, is occupied by around 10 million people which accounts for 40% of the manufacturing industry [5]. Flood risk is projected to increase across the UK with annual damages expected to increase from a current reference point of £1 billion to somewhere between £1.8 and £5.6 billion by the 2080s for England. This is with the exclusion of the effects of estimated population growth which is also a key variable to increasing flood risk [6].

WIT Transactions on The Built Environment, Vol 194, © 2020 WIT Press
www.witpress.com, ISSN 1743-3509 (on-line)
doi:10.2495/FRIAR200021

Much research has been done on flood risk, with respect to its impacts, mitigation approaches and decisions about minimising future impacts. However, most of the research carried out on buildings has concentrated almost exclusively on residential properties. There exists a dearth of research about the impacts of flooding on commercial property [7]. It is estimated that around 185,000 commercial properties are at a direct risk of being affected by flooding [8]. The statistics on the 2007 floods laid emphasis on the havoc flood can wreck on business: with records showing around 8,000 businesses were affected, resulting into 35,000 insurance claims, averaging between £75,000 and £112,000 [7]. In the 2015–2016 winter flood event, an estimate for business property damages is £513 million with a range of £410 and £616 million paid out by the insurance industry as business claims [9]. These huge sums indicate the importance of helping businesses to become more resilient and highlight the need for further research to inform practice and future policy development [7].

Commercial properties are a central component of national assets and gross domestic product (GDP) and therefore their value is of broad significance to not only the property owners but also local and national economic prosperity. Consequently, commercial property plays a significant role in the UK economy [10]. For instance, according to the Property Data Report in 2013, in the UK, the market value of the core commercial properties, such as the retail, office, hotels and industrial properties was £683 billion [11]. Commercial property also represents a major investment asset for both the pensions and insurance industries [12]. In 2010, funds held around 4.8% (or £98 billion) of their investments in direct property. Within the UK, the commercial property sector forms a significant portion of the economy with an estimated turnover of £3,200 billion and employing about 22.8 million people with an average growth rate of 2.2% since 2008 [13].

While commercial property investment may seem out-of-the-way for many people, its relevance is seen in the way rental income from business leases on offices, shops, industrial and leisure facilities supports pensions, insurance policies and savings plans [7]. Therefore, the security of these investments is paramount to the large number of shareholders and stakeholders who count on them for pensions, insurance and investment plans. Flooding can have a huge impact in hampering this security.

As means of mitigating these impacts, innovative approaches have been developed [14]. One of such is structural measures which are engineered solutions designed with defined limits of disturbance they can accommodate [15]. Once the disturbance produced is more than the specified threshold capacity of the engineered solutions, defences can be overtopped and lives and properties again become susceptible. The presence of these structural measures offers some amount of resilience to flooding, it is however very difficult to sustain and mostly overwhelmed by the next greater flood event. Despite the huge investment in structural approaches and engineering measures, flooding still remains one of the greatest threats to buildings, businesses and the wellbeing of humans. In recent years, UK flood risk management policy has recognised the need that flood cannot be totally prevented and therefore has embraced a proactive and more robust approach of managing flood risk and living with floods which is captured under the "living with water" philosophy [16]. This approach, in the context of property level flood risk, often entails building resilience into the system that is exposed to the risk. For properties, much attention has been focused towards the development and adaptation to the risk of flooding [17], [18]. This concept is commonly referred to as property level flood resilience of property flood resilience (PFR) and has, since its inception, gained wider recognition in the domain of flood risk management [14], [19].

This research draws on an examination of the literature to analyse the impacts of flooding and flood risk on commercial properties. The concept of property level flood risk mitigation is then discussed with a focus on the recoverability/resilience measures appropriate for commercial properties. The study ends with some recommendations for further research towards helping to establish this approach more widely.

## 2 FLOOD IMPACTS ON COMMERCIAL PROPERTIES

Loss and damage from the flooding of commercial properties is evident globally and seen to be prevalent in the UK [20]. Damage caused is greatly associated with the continuous interaction between the natural and human systems [21]. Flood damages, whether potential or actual, have been classified as either direct tangible, direct intangible, indirect tangible, or indirect intangible damage [22], as summarised in Table 1.

Table 1:  Classification of flood impacts on commercial properties [22].

| Flood impact | Tangible | Intangible |
|---|---|---|
| Direct | • Structural damage<br>• Damage to furniture and fittings<br>• Loss of stock<br>• Damage to equipment and machinery | • Loss of document<br>• Ill-health |
| Indirect | • Loss of production<br>• Cost of evacuation of goods and staff<br>• Clearing and cleaning up cost<br>• Repair cost | • Loss of reputation<br>• Business disruption<br>• Issues with renewing insurance |

The direct impacts entail physical damage to buildings and stock, while indirect impacts involve business disruption, lack of access and loss of business which are referred as secondary effects of flooding [20]. The tangible part of these impacts can be easily measured and claimed, like damage to building and loss of stock, while intangible impacts such as loss of reputation and issues with renewing insurance are difficult to measure and may have huge impact on a business in the long run [20]. Studies have suggested that the indirect impacts of flooding often exceed the costs of direct damage and claims for business interruption may dwarf claims against property insurance [23], [24].

### 2.1 Direct flood damage to commercial property

The damage caused as a result of direct contact with flooding relates to physical damage to business assets during a flood event. This includes damage to infrastructures, fittings and furniture, machineries, equipment and loss of stocks due to lack of mobility [25], [26]. Typically, enterprises with large fixed assets like buildings and huge inventories from raw materials to finished products are more susceptible to direct physical damage. According to Chang and Falit-Baiamonte [27], when businesses suffer from this form of damage, it can be directly linked with the total loss suffered by business.

The direct tangible impact relates to the potential cost of damage which can be estimated such as the value of physical structures or economic assets exposed to loss, while loss of business records could be classified as direct intangible impact alongside resultant ill-health of staff. However, properties with higher dealings in cash and soft business securities are safer in terms of physical vulnerability since they are intangible in nature and can be safely stored in separate locations [26]. Lost documents and records are vital physical losses and often considered intangible impacts, which can delay post event recovery work.

## 2.2  Indirect impacts to commercial property

Indirect damage is induced by flooding, but occurs, in space or time, outside the actual event [28]. Indirect losses usually result as a consequence of direct losses. The damage caused as a result of indirect contact with flooding may affect the continuity and performance of business and thereby incur loss by taking time to recover from its disrupted state of operation [29]–[31]. Therefore, even if a business escapes direct damage it may be forced to shut down as a result of indirect impacts such as disruption in supply chain, access problems for employers and employees, loss of customers and so on [30]. All expenses for disaster response, such as costs for sandbagging, evacuation and emergency services, are classified as indirect tangible damages. The cost of clearing and cleaning up and unavailability of staff (e.g. due to injuries sustained during flood event) are regarded as indirect intangible costs which can be substantial [28].

One component that can be affected by both the direct and indirect impacts is the value of commercial property. According to the RICS [32], the value is directly impacted through the physical impacts of flooding on the physical assets, while the indirect impacts are based on the social, economic and political assumptions associated with the condition of the asset at risk.

## 3  PROPERTY LEVEL FLOOD RISK MITIGATION

Whilst knowledge about the disruption and the damages caused to businesses is increasing, there is still relatively little evidence of the actions from most businesses to minimise such damages and ensure they are prepared against future risks [33]. The little evidence available is often subjective in nature and mostly concerning small and medium enterprises [33]. While traditional flood defences may be available to provide protection against coastal and river flooding for large communities, there will always be some commercial properties that would not benefit from such schemes. Such defences are not likely to deal properly with localised pluvial, surface water or groundwater flooding. Furthermore, there will always be a residual risk, as flooding cannot be totally prevented even after flood management schemes have been put in place.

The Department for Communities and Local Government [34] highlights some of the causes of this residual risk as: failure of flood management infrastructure such as a breach of a raised flood defence; blockage of a surface water sewer or failure of a pumped drainage system; a severe flood which causes a flood defence to be overtopped; and floods outside the known flood risk areas. In these cases, business owners need to have a range of protection or resilience measures they can incorporate into their properties to take care of this flood risk [35]. According to the Environment Agency, most businesses can save up to 90% on the cost of lost stock and moveable equipment by taking action to prepare in advance of flooding. Also, by preparing for flooding, they can significantly reduce financial losses; limit damage to property, stock and equipment; minimise business disruption and continuity, help to retain clients and contracts; maintain customer, supplier and business records and; obtain insurance cover [36].

In terms of protection of properties, a hierarchy of options has been recognised which is associated with decreasing residual flood risk, although this depends on the flood type and building being considered [34]. These are summarised as follows:

i.   Avoidance: comprises a range of measures including location of buildings in areas of least risk (land use planning), raising properties above the flood level, use of bunds or other hard defences to keep floodwater away.

ii.   Resistance: comprises of measures that are taken to prevent floodwater from entering into the building and damaging its fabric and contents.

iii.   Resilience: entails sustainable measures that can be integrated into the building fabric, fixtures and fittings in order to lessen the potential of damage caused by floodwater. These measures would allow for quicker drying and easier cleaning, and also ensure that the structural integrity of the building is not compromised thereby reducing the recovery time for the building to be re-occupied.

iv.   Reparability: forms a subset of resilience, covering design of elements that facilitate replacement and repair, such as sacrificial finishes.

Although property level flood risk mitigation has to a very large extent focused on residential buildings, many of the measures adopted in protecting residential properties can be applied to commercial properties. These include those designed to keep water at bay to those required to minimise floodwater impacts (both direct and indirect) when water enters into the property. These measures have been classified into two main categories, the resistance measures (also called dry proofing measures or water exclusion strategy) and the resilience measures (also referred to as wet proofing measures or water entry strategy). Table 2 shows the measures adopted in commercial properties under these two categories.

Table 2:  Categorisation of adaptation measures employed in commercial properties.

|   | Adaptation measures | Resistance | Resilience |
|---|---|---|---|
| 1 | Raised electric sockets and wirings | | √ |
| 2 | Equipment and machinery on raised plinth | | √ |
| 3 | Record back up (for customers, suppliers and staff) | | √ |
| 4 | Number of storeys | | √ |
| 5 | Emergency flood plan | | √ |
| 6 | airbrick | √ | |
| 7 | Flood guards for door and window | √ | |
| 8 | Sandbags | √ | |
| 9 | Vent covers | √ | |
| 10 | Toilet seal pans | √ | |
| 11 | Raised doors and windows | √ | |
| 12 | Sump and pump systems | | √ |
| 13 | Backup power source | | √ |
| 14 | Non-return valve on drains and pipes | √ | |
| 15 | Water resistant paint | √ | |
| 16 | Flood insurance | | √ |
| 17 | First aid kit | | √ |
| 18 | Elevators equipped with water sensor | | √ |
| 19 | Electrical panel with WIFI enabled breakers | | √ |

### 3.1  Resistance measures

The purpose of applying these kinds of measures to a commercial property is to make it watertight to floods of limited duration (a few hours) and depth (typically less than 600 mm) [34]. Consequently, this approach is often adopted up to a threshold value of 600 mm around a property, although in some cases surveyors may recommend this to be conducted up to 300 mm [37]. This will help to reduce damage to the internal fabric of the building and its contents, such as stocks, equipment and machinery, staff, customers and suppliers records, thus reducing the associated stress and suffering to business users and owners.

These measures provide property owners the opportunity to relocate important items to a safer level. In order to achieve this, the resistance approach is focused on keeping floodwater out of a building by sealing all water entry paths. It is essential that all potential entry points of floodwater are identified and protected. Any openings left unsealed serve as a passage for floodwater to enter into the building, meaning that the resistance approach fails. Work has been done to identify these potential points of water entry mainly in residential building [37], [38] and some of these findings can also be applied to commercial properties with similar features. In an ideal case, floodwater would be fully kept out of the buildings, however this may expose the building to structural risks as a result of the weight of water against walls.

Therefore, in order to adopt this kind of measure, it is essential to consider and ensure that the building structure has the capacity to resist four types of flood-related forces: (1) hydrostatic flood force that freestanding water exerts on a submerged object; (2) buoyancy force that a building receives from surrounding floodwaters; (3) hydrodynamic force that vertical surfaces receive from moving floodwaters; and (4) debris impact force to withstand the flood-borne debris strikes on the side of building [39]. The advantage of applying resistance is that the building is kept dry and the contents (stock and valuable records) inside the building are not affected by flood [37]. However, one of the disadvantages is that the stability of the building structure can be jeopardized because of the heavier load of flood water [39].

The products that make up the resistance measures include both flood protection products (such as floodgates, door and window guards for doorways and low level windows), the installation of non-return valves on sewers to prevent backflow, barriers and flood doors that cover apertures and the use of resistant materials (such as engineering bricks, cladding systems, plastic skirts, flood-resistant doors, and wall coatings to limit water ingress) [35]. However, it is recommended that above the 600 mm threshold height, a resilience approach should be adopted [34].

### 3.2  Resilience measures

A flood resilience approach involves taking measures to ensure a business can recover from the impact of a flood quickly, by minimising disruption and allowing business continuity or resumption as soon as possible. According to the Department for Communities and Local Government [34], the use of appropriate resilience measures through careful selection of building materials, construction techniques and internal finishes can help reduce the risk of flood damage to the business premise and the integral components inside the building. In terms of the building materials, these must possess properties that are resistant to flood forces, including deterioration caused by repeated inundation, and excessive moisture and humidity during and after flooding [39]. Also, because flood water may travel with sediment, chemicals and organic materials, which can be harmful to the structure and the occupants, the structure (both floor and wall) should be covered with materials that can be easily cleaned without leaving any contaminants behind [39]. Concrete, hard brick, plastic, metal, and pressure-treated wood have been identified as suitable materials for this purpose.

Meanwhile, there are variety of techniques that should be applied in order to minimise the damage caused to the building and its contents. One of these is to ensure that the building has flood vents or permanent openings that allow water to flow in and out of the structure without damaging the foundation [38]. Other measures that are vital to enabling quick recovery are to ensure that mechanical and utility facilities (such as electrical, heating, ventilation, plumbing, and air conditioning equipment) are raised above the expected flood level [37]. Vulnerable items, such as utilities, appliances, computers and contents, are relocated, permanently or temporarily, to higher parts of the building or placed above the expected flood level. Furthermore, by making changes to the location of services and electrical points such as raising plug sockets up the wall, business owners will not only improve the safety of the building for the occupants, but can also save significant amounts of money on repairing these after a flood [36].

According to the Dhonau and Rose [36], the following measures should be considered in making a commercial property resilient:

i.    raising electrical sockets, electrical wiring and controls for ventilation systems;
ii.   raising equipment and machinery on plinths;
iii.  using materials that can withstand flooding, for floors and the lower part of walls and staircases;
iv.   backing up customers' data on a regular basis;
v.    storing customers' files and supplier contracts safely;
vi.   keeping insurance policy in a secure, accessible place, as well as a copy in a "Grab bag" or "Battle box";
vii.  ensuring drains from your premises are running efficiently.

Other measures identified are retrofits such as equipping elevators with water sensors to prevent them from proceeding to flood-inundated building levels and also equipping electrical panels with WIFI enabled breakers to allow for remote shut off [40].

## 4  MOVING FORWARD

The existing literature shows that much of the research on property flood resilience (PFR) has been directed towards residential properties. Meanwhile, commercial properties represent an important component of the built environment, and very often the economic and social impact of flooding are far greater on commercial properties. There is a pressing need to investigate the application of property flood resilience (PFR) measures in the context of commercial property.

Supported by the RICS Research Trust, this on-going research project aims to investigate the application of property flood resilience (PFR) to commercial buildings. A survey of a range of commercial property owners (e.g. office, retail and industrial) located in flood prone locations is to be undertaken to examine their flood history and experience, recovery process, the availability of suitable technical interventions and, importantly, the views of the key stakeholders. The findings from this research will help to inform future flood risk policy towards the protection of commercial properties. It would also provide useful guidance to commercial property owners on how to improve their flood resilience.

This research is anticipated to produce some practical recommendations in the form of an RICS Guidance Note for commercial/general practice/building surveyors on how to implement PFR measures in commercial buildings. This will facilitate the provision of professional advice to commercial property owners and support decision making, property valuation, investment, insurance and flood risk management. The findings will be of interest and use to surveyors in all world regions. The research findings will also provide evidence-based information to inform future flood risk policy and strategy, particularly Defra and the Environment Agency. Although this research will use data from commercial properties located in the UK, the findings and recommendations are likely to be applicable to other regions and countries. The findings could be used for the basis of future research to take account of regional variations and local methods of construction.

## REFERENCES
[1]   UNISDR, *Global Assessment Report on Disaster Risk Reduction*, United Nations International Strategy for Disaster Reduction: Geneva, Switzerland, 2011.

[2]   Jha, A.K., Bloch, R. & Lamond, J., *Global Assessment Report on Disaster Risk Reduction*, The World Bank: Washington DC, 2012.

[3]   Jonkman, S., Global perspectives of loss of human life caused by floods. *Natural Hazards*, **34**, pp. 151–175, 2005.

[4]   Environment Agency, *Flood and Coastal Erosion Risk Management Long-Term Investment Scenarios (LTIS)*, Environment Agency: Bristol, 2014.

[5]   Kenney, S., Pottinger, G., Plimmer, F. & Pocock, Y., *Flood Risk and Property: Impacts on Commercial and Residential Stakeholders' Strategies*, College of Estate Management: Reading, 2006.

[6]   Committee on Climate Change, *UK Climate Change Risk Assessment 2017: Synthesis Report: Priorities for the Next Five Years*, Committee on Climate Change: London, 2016.

[7]   Pottinger, G. & Tanton, A., *Waterproof: Flood Risk and Due Diligence for Commercial Property Investment in the UK*, College of Estate Management: Reading, 2011.

[8]   Environment Agency, *Flooding in England: A National Assessment of Flood Risk*, Environment Agency: Bristol, 2009.

[9]   Environment Agency, *Estimating the Economic Costs of the 2015 to 2016 Winter Floods*, Environment Agency: Bristol, 2018.

[10]  Bhattacharya-Mis, N. & Lamond, J., Risk perception and vulnerability of value: A study in the context of commercial property sector. *International Journal of Strategic Property Management*, pp. 252–264, 2016.

[11]  British Property Foundation, *Property Data Report 2014*, British Data Foundation: London, 2014.

[12]  Investment Management Association., *Asset Management in the UK 2009–2010: The IMA Annual Survey*, Investment Management Association: London, 2010.

[13]  BIS, *SME Statistics for UK region 2008.* http://stats.bis.gov.uk/ed/sme/Stats. Accessed 22 Oct. 2019.

[14]  Oladokun, V., Proverbs, D. & Lamond, J., Measuring flood resilience: A fuzzy logic approach. *International Journal of Building Pathology and Adaptation*, **35**(5), pp. 470–487, 2017.

[15]  Proverbs, D. & Lamond, J., *Flood Resilient Construction and Adaptation of Buildings*, Oxford University Press: Oxford, 2017.

[16]  DEFRA, *Making Space for Water: Taking Forward a New Government Strategy for Flood and Coastal Erosion Risk Management in England*, Defra Publications: London, 2005.

[17]  Wingfield, J., Bell, M. & Bowker, P., *Improving the Flood Resilience of Buildings through Improved Materials, Methods and Details*, Leeds Metropolitan University: Leeds, 2005.

[18]  Kazmierczak, A. & Connelly, A., *Buildings and Flooding: A Risk-Response Case Study*, EcoCities project, University of Manchester: Manchester, 2011.

[19]  Kelly, D., Barker, M., Lamond, J., McKeown, J. & Blundell, S., *Code of Practice for Property Flood Resilience*, 1st ed., CIRIA: London, 2019.

[20]  Lamond, J. et al., *Flood Risk Mitigation and Commercial Property Advice: An International Comparison*, RICS: London, 2017.

[21]  Bhattacharya, N., A model to investigate the impact of flooding on the vulnerability of value of commercial properties. PhD thesis, University of Wolverhampton, Wolverhampton, 2014.

[22]  Merz, B., Kreibich, H., Schwarze, R. & Thieken, A., Assessment of economic flood damage. *Natural Hazards Earth Systems Science*, **10**, pp. 1697–1724, 2010.

[23]  Kleindorfer, P.R. & Germaine, H.S., Managing disruption risks in supply chains. *Production and Operations Management*, **14**(1), pp. 53–68, 2005.

[24]  Heite, M. & Merz, M., An indicator framework to assess the vulnerability of industrial sectors against indirect disaster losses. *International ISCRAM Conference*, Gothenburg, Sweden, 2009.

[25]  Tierney, K.J., Business impacts of the Northridge earthquake. *Journal of Contingencies and Crisis Management*, **5**(2), pp. 87–97, 1997.

[26]  Zhang, Y., Lindell, M.K. & Prater, C.S., Vulnerability of community businesses to environmental disasters. *Disasters*, **33**(1), pp. 38–57, 2009.

[27]  Chang, S.E. & Falit-Baiamonte, A., Disaster vulnerability of businesses in the 2001 Nisqually earthquake. *Global Environmental Change Part B: Environmental Hazards*, **4**(2–3), pp. 59–71, 2002.

[28]  Thieken, A.H. et al., Methods for the evaluation of direct and indirect flood losses. *4th International Symposium on Flood Defence: Managing Flood Risk, Reliability and Vulnerability*, Toronto, Ontario, 6–8 May 2008.

[29]  Alesch, D.J., Holly, J.N. & Nagy, R., Small business failure, survival, and recovery: Lessons from the January 1994 Northridge Earthquake. *NEHRP Conference and Workshop on Research on the Northridge*, California, 1998.

[30]  Tierney, K.J., Businesses and disasters: Vulnerability, impacts, and recovery. *Handbook of Disaster Research*, eds L.E. Quarantelli, H. Rodríguez & R.R. Dynes, Springer: New York, pp. 275–296, 2007.

[31]  Parker, J.D., Business interruption losses gauged through site surveys. *London Flood Seminar: Lighthill Risk Network*, 2009.

[32]  RICS, *RICS Valuation Standards: Global and UK*, 7th ed., RICS: Norwich, 2011.

[33]  ASC, *UK Climate Change Risk Assessment 2017 Synthesis Report: Priorities for the Next Five Years*, Adaptation Sub-Committee of the Committee on Climate Change: London, 2016.

[34]  Department for Communities and Local Government, *Improving the Flood Performance of New Buildings: Flood Resilient Construction*, Department for Communities and Local Government: London, 2007.

[35]  Tagg, A. et al., A new standard for flood resistance and resilience of buildings: New build and retrofit. *FLOODrisk 2016: 3rd European Conference on Flood Risk Management*, 2016.

[36]  Dhonau, M. & Rose, C.B., *A Business Guide to Flood Resilience*, MDA Community Flood Consultants, UK, 2016.

[37]  ODPM, *Preparing for Flood: Interim Guidance for Improving the Flood Resistance of Domestic and Small Business Properties*, Office of the Deputy Prime Minister: London, 2003.

[38]  CIRIA & Environment Agency, *Flood Protection Products: A Guide for Homeowners*, CIRIA: London, 2003.

[39]  World Meteorological Organization, *Flood Proofing: A Tool for Integrated Flood Management* (Version 1.0), WMO/GWP: Associated Programme on Flood Management, 2012.

[40]  Moudrak, N. & Feltmate, B., *Ahead of the Storm: Developing Flood-Resilience Guidance for Canada's Commercial Real Estate*, Intact Centre on Climate Adaptation, University of Waterloo, 2019.

# INTEGRATED STRATEGIES FOR RIVER RESTORATION AND LAND RE-NATURALIZATION IN URBAN AREAS: A CASE STUDY IN MILAN, ITALY

FRANCO RAIMONDI[1,2], CLAUDIA DRESTI[3], MARIANA MARCHIONI[1],
DARIO KIAN[2], STEFANO MAMBRETTI[1,4] & GIANFRANCO BECCIU[1]
[1]Politecnico di Milano, Italy
[2]Ersaf, Italy
[3]National Research Council, Water Research Institute, Italy
[4]Wessex Institute of Technology, UK

## ABSTRACT
Densely populated areas are frequently affected by floods, risking people's safety and economic activities. In Milan, Italy, the Seveso river crosses the urban area mainly in close conduits frequently flooding. The sprawling of urban areas combined with the intensification of extreme storm events increase the frequency of floods requiring pursuing a new approach on urban water management. The solutions must be sought not only on structural facilities directly on the river with large-scale dimensions: they present expensive construction and operation costs, and only give an apparent sense of security in a short period. It is necessary to identify natural-based strategies for the fluvial territory management taking a comprehensive view on watershed scale, moving from a traditional local and monothematic approach to a global and multisectoral towards water sensitive cities. This research aim is to assess some effects arising from the applications of river restoration and sustainable urban drainage techniques on a stretch of the Seveso river within Parco Nord, in particular through measures of parking de-waterproofing, improvement of river natural expansion and morphology re-naturalization and diversification of riverbanks and riverbed. To assess the effects, a 2D flow simulation using Hec-Ras and the recalculation of the river functionality index have been conducted. The results show benefits not only in raising better water and environmental quality, thanks to the enhancing of river functionality level, but also in risk mitigation, with the reduction of floodable areas, above all significantly for the storm event with a return period lower than 10 years. This research confirms the validity of the new approach and constitutes the first step towards the creation of a practical guide tool for the watershed management with similar characteristics to that of the Seveso river, to reach the European Directives' requests and to build up a strategy for adapting to climate change.
*Keywords: river restoration, spatial planning, river contract, sustainable drainage, environmental quality enhancement, risk mitigation.*

## 1 INTRODUCTION
Most of oldest cities have arisen on the water as it is a source of sustenance, a driving force and an important communication route. Over the centuries, the pursuit of progress has led to the exploitation of natural resources and the transformation of river basins, profoundly changing their characteristics over time. Floodplain occupation for new developments and the use to collect and conduct wastewater and stormwater has relegated rivers and streams to a marginal role, shifting from a resource to a liability. Watercourses pollutions pose a public health risk and floods cause people and economical loss. Traditional structural solutions not always have the expected effect, in some cases can increase the problem, specially downstream. In light of the continuous urban and climatic changes, it is necessary to start new processes for a better management of the territory which combined with the study and application of new so-called green and blue infrastructures [1] for the management of rainwater, such as those offered by sustainable drainage techniques (SUDS). They constitute the starting point for preserving people's quality of life and economic activities, by returning

WIT Transactions on The Built Environment, Vol 194, © 2020 WIT Press
www.witpress.com, ISSN 1743-3509 (on-line)
doi:10.2495/FRIAR200031

to conditions of the river basins that are as natural as possible and restoring the population-river link by rediscovering the potential and opportunities it offers [2], [3]. Due to its historically known characteristics and criticalities, the Seveso watershed lends itself well to be the subject of experimentation with these new strategies.

## 2 BACKGROUND

Seveso river's course extends (Fig. 1) for about 50 km from Como Lake to the city of Milan, which it crosses completely below the surface, starting from Ornato Street up to the beginning of the Redefossi channel on the southern border of the city, and then flowing into the Lambro river in the municipality of Melegnano. Its territory, especially the metropolitan area north of Milan, has been the scene, since the 1960s, of an intense urban development that has transformed it into one of the most economically developed but also most densely populated areas of Europe. In fact, although it represents only 2.5% (about 617 km$^2$, of which 53% is occupied by anthropic activities) of the total area of Lombardy, the watershed collects a population of approximately 2,300,000 inhabitants (more than 1/5 of the total) with a density of 3,730 ab/km$^2$. The metropolitan city expansion has increasingly reduced the river space and has led to the loss of its natural connotations, the deterioration of the water quality and the ecosystem and the increase in the flood risk.

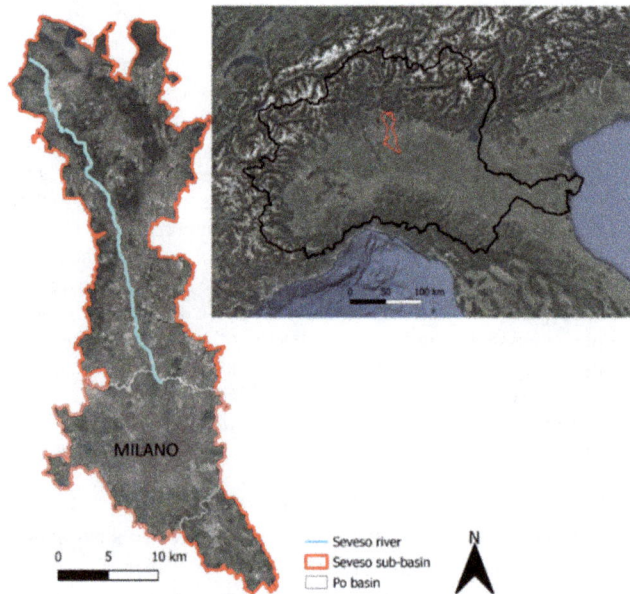

Figure 1: Seveso sub-basin location.

In particular, soil sealing has drastically reduced the percentage of rainwater infiltration and the concentration time by increasing the surface runoff. To reduce the risk of flooding, we have relied on channelization, artificialisation of the riverbed and the riverbanks and flood detention basins [4]. These solutions have not always led to satisfactory results and have often only moved the problem downstream and contributed to worsening of the water quality and the river ecosystem. It has been estimated that 104 floods have occurred in Milan since 1976 (2.6 floods each year), eight of which between 2010 and 2014. In the same way, to solve the problem of the water quality worsening, we mainly intervened on the collection network effectiveness and in an attempt to improve the yields of the treatment plants. However, the level of quality remains low as evidenced by the various quality indices (ecological, morphological and functional) which in the stretch north of Milan present the worst values such as to require an extension for the achievement of the "good" quality status imposed by the Water Directive since 2015 to 2027. It is glaring that exclusively sectoral solutions for local and structural emergency resolution are not sufficient to solve problems and sometimes present technical and economic impossibilities to achieve good results. Moreover, in the last years the rainfall regimes have changed as well as the waterproofing of the soil has increased. New solutions are necessary.

## 3  NEW APPROACH

The awareness of changing the paradigm in spatial planning has led the Lombardy Region to become the promoter of a new territorial management tool, the River Contract [5], a negotiated, voluntary and participatory planning process to which all stakeholders of a river basin can adhere. The Contract reproduces the objectives of the European Directives (mainly the Water Directive 2000/60/EC [6] and the Flood Directive 2007/60/EC [7]) and of the sector plans which implement and in turn integrate the provisions, respectively the Piano di Tutela ed Uso delle Acque (PTUA, Protection and Use Plan of the Waters) [8] and the Piano di Gestione del Rischio Alluvioni (PGRA, Flood Risk Management Plan) [9]. The limit of large area planning as well as urban and individual planning is to tackle sectoral issues and limit oneself to simple imposition or simple respect of constraints, without deeply investigating the reality of the territories. The River Contract has the specific aim of integrating and making applicable in the territory what is reported in the Directives and Plans, however taking into account the needs of the latter and planning from a watershed perspective based on the environmental characteristics. Doing so it is possible to exceed the logic administrative boundaries and to avoid the implementation of single-issue interventions, aimed exclusively at solving the site-specific problem, without evaluating the repercussions on the various environmental components and in other areas of the watershed. For this reason, within the Seveso River Contract, signed in 2006, the Strategic Sub-Basin Project has developed since 2014 and approved in December 2017, a co-planning process with local actors that leads to the development of shared project proposals aimed at mitigating the hydraulic risk and enhancing watershed quality and ecological set-up based on the interests of the stakeholders and an in-depth knowledge of the critical issues, needs and potential of the area [5].

## 4  CASE STUDY

A project idea of the redevelopment of the river environment in the north of Milan, more precisely of a stretch of about 2 km of the Seveso river within the Parco Nord, between the municipalities of Milan, Cormano and Bresso, was born. The idea, included in the action plan of the Strategic Sub-Basin Project contains a series of specific interventions aimed primarily at recovering the ecological role of the river but which simultaneously contribute to the mitigation of the hydraulic risk and the improvement of the water quality. The area presents itself as one of the most critical of the basin as regards the flood risk, due to the percentage of waterproofed surface (more than doubled from the 1950s–1960s and with values currently higher than 70% of the municipal areas involved) and the fact that it is located in the closing section of the watershed just upstream of the manhole in Ornato street in Milan, which allows the transit of an estimated flow of not more than 30–40 $m^3/s$, just sufficient for the drainage of rainwater urban areas for events that do not exceed a return time of 2 years. The risk is significantly increased by the high number of people exposed since the municipalities concerned, despite excluding Milan, have a population density of approximately 5,980 inhabitants/km$^2$. Water quality is also a critical issue since here the lowest values of the entire river rod are recorded for all quality indices. The interventions that have been purposed concerned both the river rod and the surrounding territory, as shown in Fig. 2.

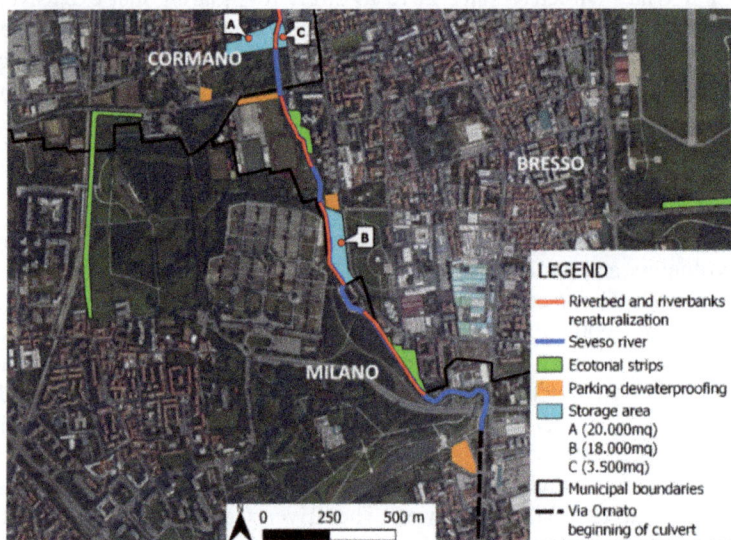

Figure 2: Proposed interventions map.

The construction of four ecotonal buffers is expected, of which two adjoining the water course for a total length of 430 meters and two others along the roads bordering the park for a total length of 1,100 m. The strips must have a width of not less than 10 m and be composed of rows of arboreal (willow, alder, poplar, oak and hornbeam) and shrubby (elderberry, privet, viburnum, hawthorn and blackthorn) species alternating. The total area covered by the ecotonal strips has been calculated in 24,000 m$^2$.

For the banks, consolidation and re-naturalization measures are envisaged. In particular, in two points the placement of cyclopean boulders at the foot of the bank surmounted by a bundle of chestnut wood with planting of shrub species suitable for the riparian environment (willow, tamarisk, privet and dogwood) or grassed through the hydroseeding technique in the case of steep slopes is expected. In a very steep stretch of about 90 m long and due to the lack of space, the introduction of gabions is expected.

In the riverbed, in correspondence with the sections of the bank in question, the aim is the positioning of triangular boulders deflectors and the creation of scrapers for the morphological diversification of the riverbed in addition to the insertion of macrophytes.

The three storage areas are to be created as follows. For area A (20,000 m$^2$ (Fig. 2)), in the municipality of Cormano, on the hydrographic right, which is currently fenced and abandoned, the wall that divides it from Seveso for a stretch of about 80 m is expected to be demolished and the replacement with fascinate alive in the first half, lower to facilitate natural expansion during floods, and of cyclopean boulders in the second half at the outer part of the loop where the hydrodynamic stresses are greater. Area C (3,500 m$^2$ (Fig. 2)) is located on the opposite side of the river in front of the first to replace an ecological platform. The construction method is identical to the previous one, after de-waterproofing, but without the use of cyclopean boulders. Area B (18,000 m² (Fig. 2)), in the municipality of Bresso, involves a public park to which the area of another ecological platform (Fig. 2) to be waterproofed. In this case the side is lowered and the foot protected by a concrete wall replaced with cyclopean and bundled boulders. In all three areas, lowering of the countryside level up to 1 m above the riverbed, the planting of indigenous shrub species and the delimitation with a grassy embankment for the protection of the external areas is expected.

The latest proposal concerns the de-waterproofing of four parking areas, that cover a total area of 13,000 m², using permeable pavements.

## 5  MATERIALS AND METHODS

For the evaluation of the effects of the hypothesized interventions, all the necessary data were collected. As regards the quality of the water, the tables were acquired which, section by section, led to the evaluation of the river functionality index (IFF) [10]. The index, developed by the Istituto Superiore per la Protezione e la Ricerca Ambientale (ISPRA (Upper Institute for Environmental and Protection Research)) [10], evaluates the functionality of the river ecosystem intended as the integration of structural, morphological and biotic factors concerning not only strictly the river but also the surrounding area. The evaluation form consists of 14 questions (including vegetation, morphology, riverbed composition, etc.) to which answers are assigned numerical values proportionally increasing with functionality. The sum of the scores obtained for each question determines the final judgment of the ecological functionality (Table 1). For Seveso, the data available date back to the beginning of 2017 and have been processed by Fondazione Lombardia per l'Ambiente (FLA (Lombardy Foundation for the Environment)) [11]. The Seveso course is characterized by decreasing values proceeding from the source to Milan. The average value corresponds to the judgement "poor", while in the case study area the average value is 55 (poor–very bad) with a maximum of 103 (middling–poor).

Table 1:  IFF index values, levels and judgements.

| IFF value | Functionality level | Functionality judgement |
|---|---|---|
| 261–300 | I | Excellent |
| 251–260 | I–II | Excellent–good |
| 201–250 | II | Good |
| 181–200 | II–III | Good–middling |
| 121–180 | III | Middling |
| 101–120 | III–IV | Middling–poor |
| 61–100 | IV | Poor |
| 51–60 | IV–V | Poor–very bad |
| 14–50 | V | Very bad |

As regards the data necessary for the hydraulic calculations, the geometric characteristics of the sections along the entire course of the Seveso obtained from the District Authority of the Po River basin (AdbPo) [12] were found, and available in shapefile and excel format from the Geoportal of Lombardy region [13]. The Digital Terrain Model (DTM) of the Lombardy region, with a 20 × 20 resolution, the shapefile relating to the intended use of the soils was obtained from the Geoportal too. From the AdbPo website [12], Annex 3 to the technical report of the project plan for the variant of the Piano di Assetto Idrogeologico (PAI

(Hydrogeological Conformation Plan)) of the Seveso stream, regarding the updating of hydrological and hydraulic analyzes, has been downloaded. In particular, the flood hydrograms upstream of the Canale Scolmatore di Nord-Ovest (CSNO (North West Filling Channel)) outlet were considered for events with return period of 10, 100 and 500 years at present. Finally, through the Agenzia Regionale per la Protezione Ambientale (ARPA (Regional Environmental Protection Agency)) [14] it has been possible to obtain the continuously measured flow rates at the hydrometric station in the municipality of Paderno Dugnano in the period 19 June 2014–5 November 2019.

The ecological impact of the action is evaluated through the analysis of the variation of the IFF index. The work consists in redefining the values to be attributed to the 14 questions in the form and the recalculation of the overall judgment for all the sections subject to intervention.

To evaluated the floodable areas instead the Hec-Ras software with mixed 1D/2D simulation was used. The geometric pattern (river and sections) was obtained by tracing the shapefiles downloaded from the geoportal in Hec-Ras which are therefore georeferenced. The Digital Terrain Model, the information layer on land cover and the satellite image were then added. The definition of Manning's roughness coefficients for each section and land use was carried out following the criteria and tables reported in the literature [15], [16].

In order to represents the beginning of the underground channel in Ornato Street, which constitutes the downstream boundary condition as the maximum flow rate is 35 m³/s, at the end of the open-air section of Seveso a culvert with an opening of $4 \times 2$ m has been inserted.

With the help of the satellite image, the outlines of the three storage areas and areas considered floodable have been defined using as boundaries the presence of road and railway embankments and including the districts that have notoriously and frequently affected by floods. Two floodable areas have been designed, one on the hydrographic left and one on the hydrographic right and south of the Ornato Street culvert. The separation between the storage areas and those considered floodable has been identified with a structure that follows the course of the ground according to the DTM by increasing the height by 1 m.

For the creation of the computational mesh, a value of 20 m per side has been set for each cell. In the presence of the three storage areas, the riverbank presents a lowering at the height of the water tie for the closest section corresponding to the flow rate which determines the achievement of the maximum drainage capacity of the culvert. In order to facilitate the complete emptying of the areas at the end of the flood event, to restore the reservoir capacity for the next event, a gate has been inserted for each area. The gate opens when the water level in the area is higher than that in the nearest section.

For each return period (10, 100 and 500 years) two simulations were carried out, one that represents the current state and one that considers the creation of the three storage areas and their simultaneous operation. As input data, the hydrograms processed by AdbPo for each return period upstream of the CSNO have been entered, decreased at each time step by 30 m³/s in the event of optimal operation of the CSNO itself in periods of flood. To these simulations are added those relating to four periods of variable length extracted from the measurements of the continuous flow rates from ARPA, also recalculated considering the flow derived by the CSNO. The periods considered coincide with particularly important events that caused flooding and damage in the northern part of the city of Milan. Specifically, the following periods were taken into consideration: 23 June–14 July 2014 and 3–27 November 2014, in which the maximum flow reached values comparable to those of events with a return period of 100 years, 31 July–8 August 2016 characterized by two peaks 5 days apart and 19 June–7 July 2018 with a maximum flow rate of 63 m³/s. Also in these cases, two simulations were carried out for each event, one in its current state and which reproduces the floods recorded in some areas of Milan and one in the scenario of complete realization and simultaneous operation of the three storage areas.

## 6 RESULTS AND DISCUSSION

From the simulations carried out using hydrograms with the return period of 10, 100 and 500 years, it emerges that the areas involved in flooding in the post operam scenario are always less than the current ones, as shown in Table 2 and as an example in Figs 3 and 4. In all three cases the flood moment is delayed by about half an hour and the time in which the most affected areas remain flooded tends to decrease. As for the height of the water in the flooded areas, this is a bit lower (5 cm) in the post operam scenario in the area on the left hydrographic, while in the remaining areas there is a substantial parity. As regards the velocity in the flooded areas, this is lower of about 0.2 m/s for the future scenarios than the present ones; the same difference is observed between the highest velocity within the riverbed, which is reached in the straight stretch downstream of the last storage area. Turning to the events actually recorded, the decrease in flooded areas is much more evident, especially in the events of June–July 2018 and July–August 2016 (Figs 5 and 6) which have the lowest maximum flow rates. In both cases, the lower speed of the current that comes out of the riverbed in future scenarios causes the water to expand very slowly. For the events of June–July 2014 (Figs 7 and 8) and the month of November 2014, having recorded flow rates comparable to those with return times of 100 years, the positive effects due to the reservoirs are less evident. The water level in the flooded areas is in general 3–5 cm lower only for 2016 and 2018 events with storage areas, while in the 2014 events is very similar. The maximum speed reached by the current is 0.2–0.3 m/s lower and the beginning of the flooding with the implementation of the interventions is delayed 30 minutes, except for November 2014. Interesting are the episodes of July–August 2016 and November 2014 in which there are two close-range peaks. In 2016, the first peak (about 48 $m^3/s$) appears to be completely absorbed by the reservoir system while in 2014 this does not occur and this is due to the exceptional nature of the period which has flow rates almost constantly above 35 $m^3/s$ with peaks of 73 and 120 $m^3/s$ within 3 days, a sign that the system tends to be ineffective when the maximum conveying capacity of the culvert is equalled or exceeded, albeit slightly, for a few days consecutively.

Table 2:  Expansion of flooded areas.

| Event | Time | Floodable areas ($km^2$) | | |
|---|---|---|---|---|
| | | Current state | Storage area | Gap |
| 10 years return period | 5 hours after the beginning | 5.1 | 4.3 | 0.8 |
| 100 years return period | 5 hours after the beginning | 6.3 | 5.5 | 0.8 |
| 500 years return period | 5 hours after the beginning | 8.4 | 6.2 | 2.2 |
| 23 June–14 July 2014 | 8 July 2014, 05:30 | 6.3 | 5.4 | 0.9 |
| 3–27 November 2014 | 15 November 2014, 20:30 | 8.5 | 8 | 0.5 |
| 31 July–8 August 2016 | 7 August 2016, 00:00 | 3.2 | 1.3 | 1.9 |
| 19 June–7 July 2018 | 5 July 2018, 08:00 | 2.4 | 1.1 | 1.3 |

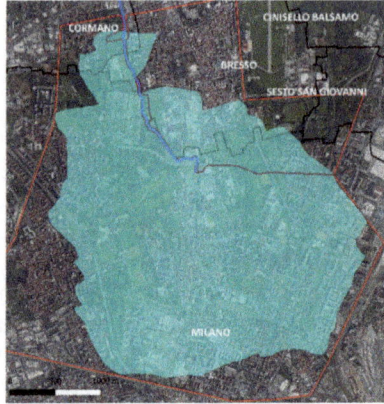

Figure 3:  Flooded areas during 500 years return period event at the current state.

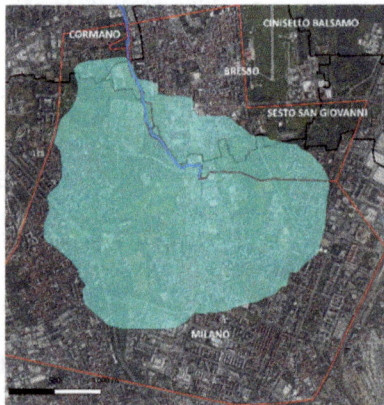

Figure 4:  Flooded areas during 500 years return period with storage areas.

Figure 5:  Flooded areas on 7 August 2016 (00:00).

Figure 6: Flooded areas on 7 August 2016 (00:00) with storage areas.

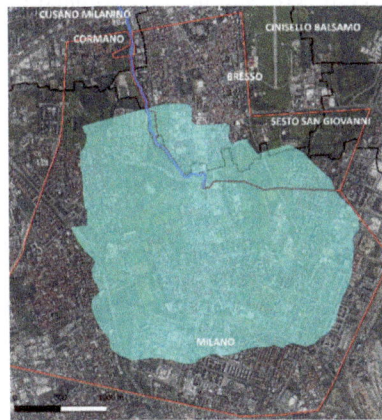

Figure 7: Flooded areas on 8 July 2014 (05:30).

Figure 8: Flooded areas on 8 July 2014 (05:30) with storage areas.

For the ecological aspect, the evaluation of the hypothesized interventions is entrusted to the recalculation of the river functionality index. Based on the location of the re-naturalization and morphological diversification of the riverbed and riverbanks, five files have been reviewed, three concerning the hydrographic right and two, the left. Everywhere, the implementation of the project proposals determines the transition to the next level of functionality of the IFF, with an increasing of value between 32 and 63 points. In particular, the improvements enhance the functionality judgement from poor to middling in three cases and from poor–very bad to middling–poor and from very bad to poor in the other two. Among the various questions that make-up the index, those that benefit most from the proposed actions and that determine the improvement of functionality are the quality, continuity and width of the vegetation in the primary and secondary peripheral areas, the conformation of the riverbed, the stabilization of the riverbanks and in some cases the enlargement and re-naturalization of the cross sections. It is very difficult to be able to reach better judgments due to the serious state of neglect facing the river at this time and the lack of physical space to allow to operate more consistently.

In addition to the tangible results of the IFF and the hydraulic simulations, there are a whole series of positive effects directly and indirectly linked to the proposed interventions that are not easily quantifiable or in any case not in the short term and which concern both the improvement of water quality and ecological functionality of the river and the decrease in hydraulic risk. From a water quality point of view, some interventions, although not acting directly for this purpose, are still able to lighten the polluting load linked to the washout of the cemented surfaces and increase the self-cleaning capacity of the river. To the first purpose contributes the de-waterproofing of the car parks; several studies [17], [18] demonstrate how the coating of parking surfaces with techniques that facilitate the infiltration of rainwater determines a reduction of the polluting load especially related to suspended solids and in some cases also to some heavy metals [19] being trapped in the pervious asphalt without soil contamination. Moreover, it is well known that pervious pavements have the capability to delay runoff reducing peak discharge [20], [21]. The increase in self-purifying capacity is guaranteed instead from the morphological diversification of the riverbed such as the insertion of scrapers, boulders and brushes which increase the turbulence locally and contribute to lowering the water temperature and improving oxygenation, aspects useful for removing of excess nutrients. Placement of boulders has a positive effect in improving coarse particulate organic matter retention too [22]. In addition, these solutions, if well designed, allow the current to be conveyed to the center of the riverbed during lean periods, avoiding the formation of stagnant pools and therefore of poorly hygienic conditions. It has been demonstrated that the diversification of the riverbed allows the formation of fundamental microhabitats [23] for the restoration of macroinvertebrates, that perform several functions in river ecosystem [24], and for a future return of fishes. Moreover, the presence of macrophyte species enhance photosynthetic productivity [25] and creates favorable conditions for the development of communities of microorganisms that through various types of chemical–physical processes are able to purify the course of pollutants as well as contribute to the sediment retention [26].

In general, the erosive problem of undermining the foot of the riverbanks is instead solved by the naturalistic engineering and morphological diversification interventions for which the probability of trees falling in the riverbed is considerably reduced.

Finally, the ecotonal strips on the roads allow a pre-treatment of rainwater while beyond the riverbanks they contribute to stabilize them and offer refuge and sustenance for the numerous species of birds and small mammals that live in the park, partially giving back to the river the role of ecological rod which belongs to him.

## 7 CONCLUSIONS

The proposal is a good example of a search for shared and integrated solutions to critical issues in the river sector and is evidence of the awareness of the need to make a change in the approach to spatial planning and in the effectiveness of the co-planning process with the local actors identified by the Lombardy Region in the Sub-basin Strategic Project. From an ecological point of view the actions proposed are able to enhance river environment and consequently water quality also if at a local scale. Regarding risk mitigation the system seems to be able to significantly reduce the floodable areas for events with return period lower than 10 years, while the benefits during extreme events are limited. This proposal is the starting point towards the restoration of a situation that is as close as possible to naturalness. However, in this strongly anthropized context, it is necessary not to remain focused only on the river auction and the surrounding areas, but to assess the possibility of intervention throughout the watershed. In this sense, it is necessary to carry out an overall evaluation of the entire hydrological balance, also in light of the hydraulic invariance regulation [27], and to promote flexible and sustainable solutions for the accumulation, treatment [28] and reuse of rainwater by promoting infiltration in the soil and increasing the concentration time to obtain sensitive results on a watershed scale from all points of view (ecological, water quality and risk mitigation) and as a strategy for adapting to climate change. Finally, the decision support offered by the scientific models that confirm the effectiveness of the process and help progress towards the change of approach to planning that is required cannot be ignored.

## REFERENCES

[1]   Lawson, E. et al., Delivering and evaluating the multiple flood risk benefits in blue-green cities: An interdisciplinary approach. *WIT Transactions on Ecology and the Environment*, vol. 184, WIT Press: Southampton and Boston, pp. 113–124, 2014.

[2]   Owens-Viani, L., Restoring urban streams offers social, environmental and economic benefits. *Sustainable Use of Water*. California Success Stories, Pacific Institute: Oakland, California, pp. 283–304, 1999.

[3]   Everard, M. & Moggridge, H.L., Rediscovering the value of urban rivers. *Urban Ecosystems,* **15**(2), pp. 293–314, 2012. https://doi.org/ 10.1007/s11252-011-0174-7.

[4]   Becciu, G., Ghia, M. & Mambretti, S., A century of works on River Seveso: From unregulated development to basin reclamation. *International Journal of Environmental Impacts*, **1**(4), pp. 461–472, 2018. https://doi.org/10.2495/EI-V1-N4-461-472.

[5]   Regione Lombardia, ERSAF, Progetto Strategico di Sottobacino del torrente Seveso, 2017. (In Italian.) www.contrattidifiume.it.

[6]   European Union, Directive 2000/60/CE of the European Parliament and of the Council Establishing a Framework for the Community Action in the Field of Water Policy.

[7]   European Union, Directive 2007/60/CE of the European Parliament and of the Council on the Assessment and Management of Flood Risk.

[8]   Regione Lombardia, Unità Organizzativa Risorse Idriche e Programmazione Ambientale, Piano di Tutela e Uso delle Acque, 2016. (In Italian.) www.regione.lombardia.it.

[9]   Autorità di bacino distrettuale del fiume Po, Piano di Gestione del Rischio Alluvioni, 2015. (In Italian.) www.adbpo.gov.it.

[10]  ISPRA, IFF 2007-indice di funzionalità fluviale, 2007. (In Italian.) www.isprambiente.gov.it.

[11]  Fondazione Lombardia per l'Ambiente, www.flanet.org.

[12]  Autorità di bacino distrettuale del fiume Po, Allegato 3 alla relazione tecnica del progetto di variante al pai del torrente Seveso, 2017. (In Italian.) www.adbpo.gov.it.
[13]  Geoportale Regione Lombardia, www.geoportale.regione.lombardia.it/download-dati.
[14]  Agenzia Regionale per la Protezione Ambientale, www.arpalombardia.it.
[15]  Chow, V.T., *Open Channel Hydraulics*, McGraw-Hill: New York, pp. 101–125, 1994.
[16]  Cowan W.L., Estimating hydraulic roughness coefficients. *Agricultural Engineering*, **37**(7), 1956.
[17]  Marchioni, M.L. & Becciu, G., Permeable pavement used on sustainable drainage systems (SUDs): a synthetic review of recent literature. *WIT Transactions on The Built Environment*, vol. 139, WIT Press: Southampton and Boston, p. 12, 2014. http://dx.doi.org/10.2495/uw140161.
[18]  Pratt, C.J., Permeable pavements for stormwater quality enhancement. *Urban Stormwater Quality Enhancement: Source Control, Retrofitting, and Combined Sewer Technology*, ASCE, 1990.
[19]  Legret, M. & Colandini, V., Effects of a porous pavement with reservoir structure on runoff water: Water quality and fate of heavy metals. *Water Science and Technology*, **39**(2), pp. 111–117, 1999. http://dx.doi.org/10.1016/s0273-1223(99)00014-1.
[20]  Marchioni M. & Becciu G., Experimental results on permeable pavements in urban areas: A synthetic review. *International Journal of Sustainable Development and Planning*, **10**(6), pp. 806–817, 2015.
[21]  Pratt, C.J., Mantle, J. & Schofield, P., UK research into the performance of permeable pavement, reservoir structures in controlling stormwater discharge quantity and quality. *Water Science and Technology*, **32**(1), pp. 63–69, 1995. https://doi.org/10.1016/0273-1223(95)00539-y.
[22]  Lepori, F., Palm, D. & Malmqvist, B., Effects of stream restoration on ecosystem functioning: Detritus retentiveness and decomposition. *Journal of Applied Ecology*, **42**, pp. 228–238, 2005. https://doi.org/10.1111/j.1365-2664.2004.00965.x.
[23]  Groll, M., Relations between the microscale riverbed morphology and the macrozoobenthos: Implications for the ecological quality assessment and the definition of reference conditions. *International Journal of Environmental Impacts*, **1**(3), pp. 375–389, 2018. https://doi.org/10.2495/EI-V1-N3-375-389.
[24]  Wallace, J.B. & Webster, J.R., The role of macroinvertebrates in stream ecosystem function. *Annual Review of Entomology*, **41**, pp. 115–139, 1996. https://doi.org/10.1146/annurev.en.41.010196.000555.
[25]  Jarvie, H.P., Love, A.J., Williams, R.J. & Neal, C., Measuring in-stream productivity: The potential of continuous chlorophyll and dissolved oxygen monitoring for assessing the ecological status of surface waters. *Water Science and Technology*, **48**(10), pp. 191–198, 2003.
[26]  Schmid, B.H., Innocenti, I. & Sanfilippo, U., Characterizing solute transport with transient storage across a range of experiments in Austrian and Italian streams. *Advances in Water Resources*, **33**, pp. 1340–1346, 2010.
[27]  Regione Lombardia, Regolamento recante criteri e metodi per il rispetto del principio dell'invarianza idraulica ed idrologica, 2017. (In Italian.) www.regione.lombardia.it.
[28]  Becciu, G. & Raimondi, A., Probabilistic analysis of the retention time in stormwater detention facilities. *Procedia Engineering*, **119**, pp. 1299–1307, 2015.

# POLICY DESIGN IN FLOOD RISK MANAGEMENT: STUDYING POLICY PREFERENCES IN THREE SUB-CATCHMENT AREAS IN SWITZERLAND

ANIK GLAUS
Institute of Political Science and Oeschger Centre for Climate Change Research,
University of Bern, Switzerland

## ABSTRACT
Floods are an extensive environmental problem that touches upon several policy sectors, decision-making levels, and territories simultaneously. To effectively cope with floods, decision-makers increasingly need to consider cross-sectoral, multi-level, and transterritorial solutions. However, such boundary-spanning policy solutions converge with traditional sector-, level-, and territory-oriented flood risk management. Overcoming these particular interests and moving towards a more integrated approach is therefore a complex task. However, combining policies from different entities into an integrated approach, actors' policy preferences for single instruments or instrument mixes are key. With the aim of understanding effective policy design in flood risk management, this paper studies whether actors' preferences for flood risk management instruments align. We take the ideal case of Swiss flood risk management and analyze three hydrological sub-catchments of the Aare, Kander and Thur rivers. We surveyed and interviewed public and private actors involved in flood risk management belonging to multiple sectors, levels, and territories on their preferred instruments and instrument mixes. Based on these preference data, we evaluated the effectiveness of flood risk management measures and measure mixes via an index. Results suggest that actors' focus on traditional sector-, level-, and territory-oriented flood risk management policies outweighs preferences for more integrated approaches in Switzerland.
*Keywords: policy design, instrument mix, policy preferences, flood risk management.*

## 1 INTRODUCTION
Flooding offers an ideal example for studying policy design concerning complex environmental problems in complex multi-dimensional policy settings. Increasing frequencies and magnitudes of floods and growing flood damages in Europe pose a high risk for population and infrastructure. The cross-sectoral, multi-level, and transterritorial nature of flooding calls for effective policy design that can exploit synergies between different sectors, levels, and territories. Traditional flood risk management is, however, often organized in sectoral, political, and territorial "silos". Overcoming these particular interests and moving towards coordinated and boundary-spanning policies, known as integrated flood risk management, is a complex task and often lacks political support. Complex environmental problems such as floods that are characterized by a mismatch between affected sectors, levels, and territories require policy solutions capable of connecting these disentangled parts [1]. Some concepts in the literature, such as collaborative or polycentric governance [2], or functional spaces [3], suggest approaching complex environmental problems with an effective instrument mix that addresses public and private actors belonging to multiple sectors, levels, and territories simultaneously. Policy designs including an effective instrument mix rather than single policy instruments can fulfil various goals, interests, and priorities, address numerous challenges, and reach multiple actors. In designing an effective instrument mix, it is crucial to understand the context in which an instrument mix applies, i.e. the plurality of actors' norms, values, and interests, which can lead to a variety of often divergent preferences for different solutions [4]. Thus, previously existing arrangements,

WIT Transactions on The Built Environment, Vol 194, © 2020 WIT Press
www.witpress.com, ISSN 1743-3509 (on-line)
doi:10.2495/FRIAR200041

actor constellations, and long-standing preferences in a particular setting, influence actors' instrument choice [5]. Research on policy preferences for effective instrument mixes, however, is still limited. Consequentially, this study poses the following research question: *How do multiple actors' preferences for an effective instrument mix vary between sectors, levels, and territories?*

Addressing this research question helps to provide insight into and evaluate multiple actors' divergent preferences for different policies. An instrument mix only has the potential to effectively manage cross-sectoral, multi-level, and transterritorial complex environmental problems when supported and preferred by actors belonging to the affected sectors, levels, and territories. Following the literature, it is difficult to arrive at a common perspective between a multitude of actors with significantly different interests [1]. In this vein, the study is based on the assumption that integrated flood risk management is only possible if multiple actors belonging to different sectors, levels, and territories participate in policy design as their preferences are driven by their specific "silo".

Empirically, the study analyzes flood risk management in three hydrological sub-catchment areas of the Aare, Kander, and Thur rivers in Switzerland. Public and private actors belonging to different sectors and levels are surveyed on their preferred instrument mix in flood risk management. Based on this data, the effectiveness of instrument mixes, operationalized by the number (*density*), coerciveness (*intensity*) [6], and inclusiveness (*balance*) of instruments, in Swiss flood risk management are evaluated. By connecting the two well-known criteria *density* and *intensity*, and adding a third new indicator *balance*, an "Effective Policy Mix Index" is constructed to compare preferred instrument mixes between multiple actors. In combination with the index, qualitative in-depth interviews with key actors involved in flood risk management were conducted to contextualize preferences for particular instrument portfolios.

## 2 THEORETICAL FRAMEWORK

### 2.1 Policy design: instruments, mixes, and preferences

Policy design, and particularly instrument selection, is an inherent part of the policy formulation process [5]. Designing a policy implies that goals and targets are defined and connected to instruments expected to achieve the defined goals. As such, a policy attempts to alter aspects of social behavior and alleviate an underlying societal problem [7]. In particular, the instrumental orientation of modern policy design studies is central. For instance, a broad literature on instruments emerged, which can be delineated between an "old" school of traditional instrument studies and a "new" school of policy design orientation [8].

The "old" school of instrument studies analyzed different kinds of instruments, their characterization, and into which instrument types these instruments could be categorized (e.g. based on state action and government resources [9], degree of state intervention [10], or policy targets and their behavior [11]). Of note is that the "old" school of instrument studies was often criticized for its focus on single instruments. Governments often adopt multiple instruments in a policy field and bundle them in policy programs or instrument portfolios [12]. Therefore, the "new" school of policy design orientation began to assess more complex policy mixes including multiple instruments and identify complementarities and conflicts within instrument mixes [13]. Often new mixes are constrained by the instrument choices that have become institutionalized previously in a policy field. Policy makers choose and combine some specific instruments rather arbitrarily, especially those already well known from other contexts. In addition, governments seldom abolish existing instruments, and instead often introduce new instruments on top of existing ones.

The "new" policy design studies' approach provides valuable insights into the design of more complex forms of policies in challenging contexts, such as complex environmental problems. At the same time, however, in designing and selecting a well-functioning, productive instrument mix in a complex policy design process, it is of major importance to understand policy makers' preferences for particular instruments in a specific context [14]. Thus, previously existing arrangements, long-standing preferences, or the political context within which policy makers operate can shape the design of an instrument mix [5].

## 2.2  Evaluating design features of policy portfolios

To select from a wide range of instruments for a mix, instruments are often compared in terms of their effectiveness. Evaluating the effectiveness of instrument mixes or policy portfolios across time, policy fields or regions, many policy design studies use the two dimensions *density* and *intensity* [6]. *Density* explores the number of policies or instruments that are applied within a policy field over time [15]. Meanwhile, to account for the content of instruments [16], *intensity* provides information about the level of regulatory standards (e.g. emission limits) and their scope of application (e.g. specific branches).

This study aims to construct an index ("Effective Policy Mix Index"), which also builds on Knill et al. [6], [15] *density* and *intensity* dimensions, while adding a third dimension of instrument *balance* (see [17]). *Balance* facilitates the assessment of actors' preferences for a balanced instrument mix, including tools of all instrument types available in a particular policy field. Thus, the index does not exclusively evaluate the number and coerciveness of instruments, but additionally indicates the inclusiveness or the extent to which actors are willing to support a mix's instrument type *balance*, i.e. instruments representing multiple actors' preferences. Consequentially, the probability of eventually adopting and implementing an effective instrument mix in a complex context increases [17].

## 2.3  Actor-centered hypotheses

According to Landry and Varone [18], three groups of participants should be considered in a policy design process. First, policy makers (e.g. elected representatives), with re-election as their ultimate goal, are interested in formulating flexible policy designs, because they must react to citizens' changing preferences (i.e. policy responsiveness). Second, policy implementers (e.g. administrative agencies) prefer policy designs that maximize their financial resources and decision-making powers. Finally, target groups seek to influence policy designs in order to minimize their costs and maximize their benefits that come along with the introduced instruments. Because actors have traditionally not been equally affected by policy design processes, this differentiation of participants in three groups is key [19]. Some actors ("sources") are identified as causing a particular problem (e.g. water pollution), while other actors ("victims") are negatively affected by the current problem [20]. "Victim" actors wish to be compensated and see the problem (re)solved with the most effective solution, i.e. tend to prefer an instrument mix consisting of multiple, coercive, and inclusive instruments that targets a problem successfully. The "source" actors, however, whose stance is threatened by the introduced policies, wish to keep their costs low and their flexibility high [18], while attempting to deflect responsibility for the problem, i.e. prefer an instrument mix consisting of few, less coercive and less inclusive instruments influencing them only minimally or not at all. Consequently, depending on actors' affectedness and role in a policy design process, their demands concerning an effective instrument mix differ.

Actors in a policy design process tend to agree on similar policy designs when they share similar core beliefs regarding a specific issue (according to the *Advocacy Coalition Framework (ACF)* [4]). For instance, similar beliefs can be expressed in shared fundamental norms and values, but also in preferences for the same instruments in approaching a specific problem [19]. On the fundament of similar core beliefs, actors form coalitions in a policy process. In particular, actors facing an overlapping problem tend to work collaboratively in addressing their mutual hindrance and hence share similar beliefs in its resolution. Such shared beliefs are often the case for actors belonging to the same policy sector (or policy sub-system in *ACF*-language) addressing a particular issue from a similar perspective. Since complex environmental problems often create "victims", i.e. directly affected sectors, actors tend to agree on similar coordinated and effective policy designs. Actors belonging to sectors, which are less directly affected by a problem, however, tend to favor non-coordinated single instruments.

> **H1a**: *Actors belonging to a policy sector being directly affected by a problem tend to prefer an effective instrument mix.*

> **H1b**: *Actors belonging to a policy sector being indirectly affected by a problem tend to reject an effective instrument mix.*

As is the case with actors belonging to the same sector, actors belonging to the same decision-making level tend to agree on similar policy designs in a policy process [21]. In decentralized political systems, such as Switzerland, the national, sub-national, and local government levels often share policy competences and tasks in order to effectively address various complex problems. Despite adopting a multi-level lens, such shared policies treat different aspects of a problem at each of the levels individually. Separate problem handling leads to the development of a common sense of problem understanding, fosters collective action, and supports learning processes at each level [22]. This level-specific understanding of a current problem contributes to shared beliefs and agreements on similar policy designs [21]. Since complex environmental problems mostly manifest on the local level, local actors tend to agree on similar coordinated and effective policy designs. National and sub-national actors, however, positioned further away from a problem's consequences, are only indirectly affected by the problem, and therefore tend to favor non-coordinated single instruments.

> **H2a**: *Actors belonging to a decision-making level being directly affected by a problem tend to prefer an effective instrument mix.*

> **H2b**: *Actors belonging to a decision-making level being indirectly affected by a problem tend to reject an effective instrument mix.*

## 3 CASE STUDY

With its geographic position at the source of several major European rivers, many small-sized and densely populated areas, and increasing climate change impacts, some Swiss regions are heavily exposed to flood risks. This historical record explains Switzerland's long experience with flood risk management and a wide range of different flood-related policies and instruments. Swiss flood risk management has traditionally been characterized by an infrastructure-oriented regime, slowly shifting towards a more nature-oriented and sustainable spatial planning approach, including new integrative and coordinated risk management elements. To this day, technical instruments remain the most widespread instrument type [23]. In complex settings, however, where flooding potentially affects actors belonging to multiple sectors, levels, and territories, the demand for more integrative, coordinated, and boundary-spanning instrument types increases: technical instruments are completed by spatial planning, ecological river restoration, and information [24].

The study analyzes actors' preferences in the case of flood risk management in three hydrological sub-catchment areas in the river basins of the Aare, Kander, and Thur in Switzerland. A postal mixed-mode survey based on standardized questions is designed to gather data on actors' instrument preferences in the three sub-catchment areas. Additionally, 21 key actors are interviewed in semi-structured interviews. To identify the central actors of the three flood risk management processes, the commonly used (in the social sciences) decisional, positional, and reputational approaches are applied. The actor sample includes representatives from the federal, cantonal, and municipal administration, regional associations, interest groups, such as nature conservation organizations and leisure clubs, and economic, infrastructure, and scientific actors. 206 actors are surveyed in total (82 in the Aare, 63 in the Kander, and 61 in the Thur sub-catchment) of which 142 or 69% responded.

## 4 METHOD

### 4.1 Operationalization of "Effective Policy Mix Index"

To construct an index capturing actors' preferences for an effective policy mix, three dimensions of an instrument mix are measured: *density*, i.e. the number of instruments, *intensity*, i.e. the coerciveness of instruments, and *balance*, i.e. the inclusiveness of instruments. *Density, intensity*, and *balance* operationalizations are based on a survey question measuring actors' preferences for different flood risk management instruments. For each item in a statement battery, two single instruments belonging to different instrument types (technical, spatial planning, ecological, informative) are contrasted. The actors express their preferences for one or the other instrument on a two-dimensional four-point Likert scale ranging from full or partial agreement for one instrument to full or partial agreement for the other instrument. In so doing, actors assign to each of the two flood risk management instruments a level of preference between 1 (weak) and 4 (strong).

This preferences data provides the basis for the construction of the index indicators *density, intensity*, and *balance*. The three indicators are combined in a multiplicative index. Thus, the higher the number, coerciveness, and inclusiveness of their preferred instruments, the higher actors' preferences for an effective policy mix.

As is true for many empirical studies preceding this one, the indicator *density* is measured by counting the number of preferred instruments. An instrument is counted when actors assign at least a preference level of 3 or 4 (partial or full agreement). The number of preferred instruments is summarized for each actor and lies between 0 and 12 (Kander and Thur sub-catchments), and 0 and 10 (Aare sub-catchment), respectively. Finally, the values of the indicator *density* are normalized to a range from 0 to 1.

The indicator *intensity* can be measured empirically in various ways. In this study, the level of state action and resources available to public authorities, i.e. the coerciveness of the preferred instruments, is crucial. Hood's categorization distinguishes between *nodality*, *organization*, *treasure*, and *authority*, with increasing coerciveness from the first to the last [9]. Henstra [25] adjusts this categorization to climate adaptation instruments, which sets the foundation for the coerciveness evaluation of actors' preferred flood risk management instruments in this study. First, each instrument is assigned to a coerciveness category from *nodality* to *authority* (where *treasure*, for reasons of effectiveness, is divided into *ecosystem management* and *public goods and services*). Subsequently, for every actor, the mean value of preferences per coerciveness category is calculated. Next, these average preferences values are weighted from 1 to 5, where the least coercive category (= *nodality*) receives a value of 1 and the most coercive category (= *authority*) a value of 5. This weighting process corresponds to the assumption that directly affected actors prefer instruments that are more coercive. Finally, the values of the indicator *intensity* are normalized to a range from 0 to 1.

The indicator *balance* uses the multiple combinations of instrument types resulting from the survey question. First, every combination of the four instrument types – technical, spatial planning, ecological and informative instruments – is represented at least twice in the survey question (with different contrasted single instruments, however). Actors' preferences for the same instrument type over the other in both combinations are evaluated, to see whether they hold consistent preferences. Second, the number of instrument types in which actors have consistent preferences in both combinations is summarized. The higher the number of instrument types where actors show consistent preferences, the more they prefer a balanced mix of different instrument types. Holding preferences for all four instrument types is the maximum value and corresponds to actors' preferences for a fully balanced mix of instrument types. Finally, the values of the indicator *balance* are normalized to a range from 0 to 1.

## 4.2 Actors

For the *policy sector*, actors in Swiss flood risk management can be distinguished in seven different sectoral groups, three of which are water-related sectors – water use, protection of water, flood protection – and four of which are external sectors – cities and municipalities, agriculture and forestry, economy and infrastructure, science. Distinguishing between the water-related sectors is of utmost importance, because often the goals of using water (e.g. drinking water), protecting water (e.g. wastewater treatment), and flood protection (e.g. infrastructure construction) conflict with each other. For *actor level,* actors are distinguished according to whether they belong to the national, cantonal, regional, or local decision-making level. Actors from the three sub-catchment areas are combined into one larger sample; however, their respective sub-catchments serve as control variables.

## 4.3 Method of data analysis

First, actors' index results are analyzed univariately and compared between sectors, levels, and sub-catchments. In the bivariate analysis, Spearman's rank order correlation and Cronbach's alpha are computed for the three index indicators. Second, index results are complemented by insights gained through several in-depth interviews with key actors. These interviews provide the study with the necessary case knowledge, which helps to evaluate and interpret the index's and the three indicators' results. As such, this study adopts a mixed-mode method, combining quantitative and qualitative preferences data.

## 5 RESULTS

### 5.1 Descriptive analysis of "Effective Policy Mix Index"

The mean value for the index is 0.17, which is low and indicates that the surveyed actors generally prefer non-coordinated single instruments to manage flood risks. However, examining the indicators *density*, *intensity* and *balance* individually relativizes the impression that actors reject all aspects of an effective instrument mix. Instead, actors seem to prefer a mix including a medium number of instruments (around 5–6 instruments), with mid-coercion (some technical and/or spatial planning instruments combined with some ecological and/or informative instruments), and belonging to a balanced number of instrument types (around 2–3 instrument types).

Spearman's rank order correlation analysis of the three indicators reveals that *density*, *intensity*, and *balance* are significantly and positively correlated to each other (*density-intensity*: 0.64/*density-balance*: 0.36/*intensity-balance*: 0.40). The Cronbach's alpha measuring the reliability of the index and the importance of each indicator for the index is consistent with the Spearman's rank order correlation: 0.77 (CI: 0.70, 0.83), i.e. moderate to high, signifying that the indicators *density*, *intensity*, and *balance* are linked and can be combined in an index.

## 5.2 "Effective Policy Mix Index" by sector and level

Fig. 1 illustrates actors' preferences by sector and level. First, it is of note that the less directly affected sectors water use (cantonal and regional = 0.20) and water protection (national = 0.19/cantonal = 0.20/regional = 0.16) carry on average higher preferences for the index than the directly affected flood protection sector (national = 0.07/cantonal = 0.24/regional and local = 0.15). While actors in the water use and water protection sectors show homogeneous preferences for the different levels, in the flood protection sector, no homogeneous preferences can be found: national actors have low preferences for a mix, while cantonal actors in comparison show higher preferences. Regional and local actors have medium preferences. Second, the same situation can be observed for the agriculture and forestry sector (national = 0.12/cantonal = 0.18/regional = 0.10), a water-external sector strongly affected by flood risk management instruments to be implemented. Cantonal actors have slightly higher preferences for a mix than actors on the national or regional level.

Fig. 2 illustrates actors' preferences by sectors and sub-catchments. Results are consistent: actors from the flood protection and the agriculture and forestry sectors show heterogeneous preferences for a mix in comparison with the other sectors displaying more homogeneous preferences. For the flood protection sector, actors in the Aare sub-catchment (0.24) show higher preferences than actors in the Kander and Thur sub-catchments (both 0.14). Similarly, for the agriculture and forestry sector, actors in the Thur sub-catchment (0.22) express higher preferences for a mix than do actors from the Aare and Kander sub-catchments (both 0.07).

## 5.3 Interviews

Several actors emphasize in the interviews that effective flood risk management requires a combination of multiple equivalent instruments. However, an optimal instrument mix depends on the location and the technical options, and may change frequently. Regarding the instrument types, almost half of the actors state that spatial planning or ecological instruments should be given priority, but that the traditional technical instruments are implemented more quickly and in a more practical way. These statements allow for embedding the index's descriptive results: the overall index results are low, i.e. in general, actors prefer non-coordinated single instruments. Nonetheless, the *balance* of different instrument types appears an important dimension in the mix. Technical instruments remain the most important instrument type, but actors state that they prefer to combine them with spatial planning and/or ecological instruments. Thus, actors wish to combine at least 2–3 instrument types rather than rely only on technical instruments.

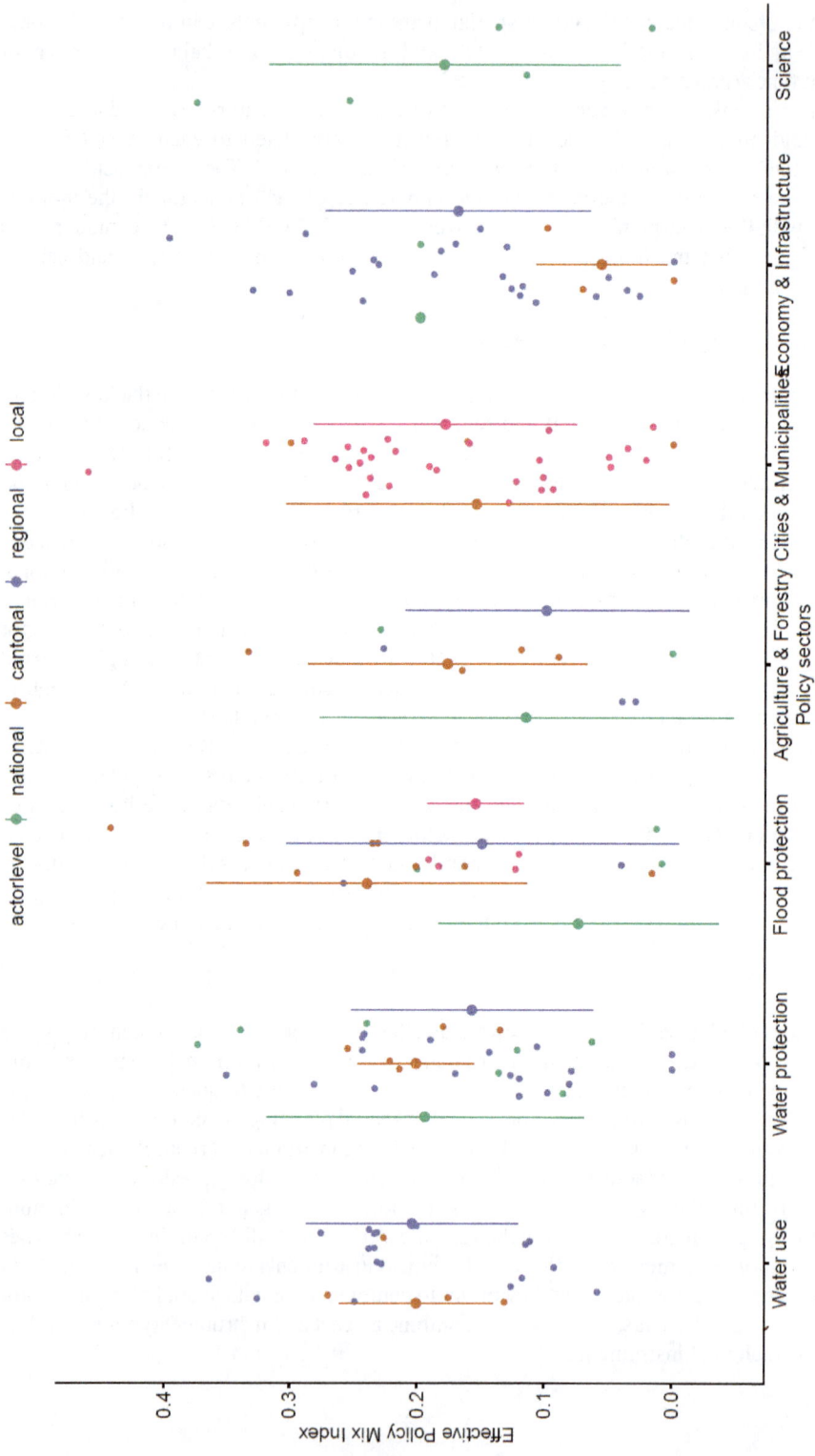

Figure 1: Actors' preferences by sector and level.

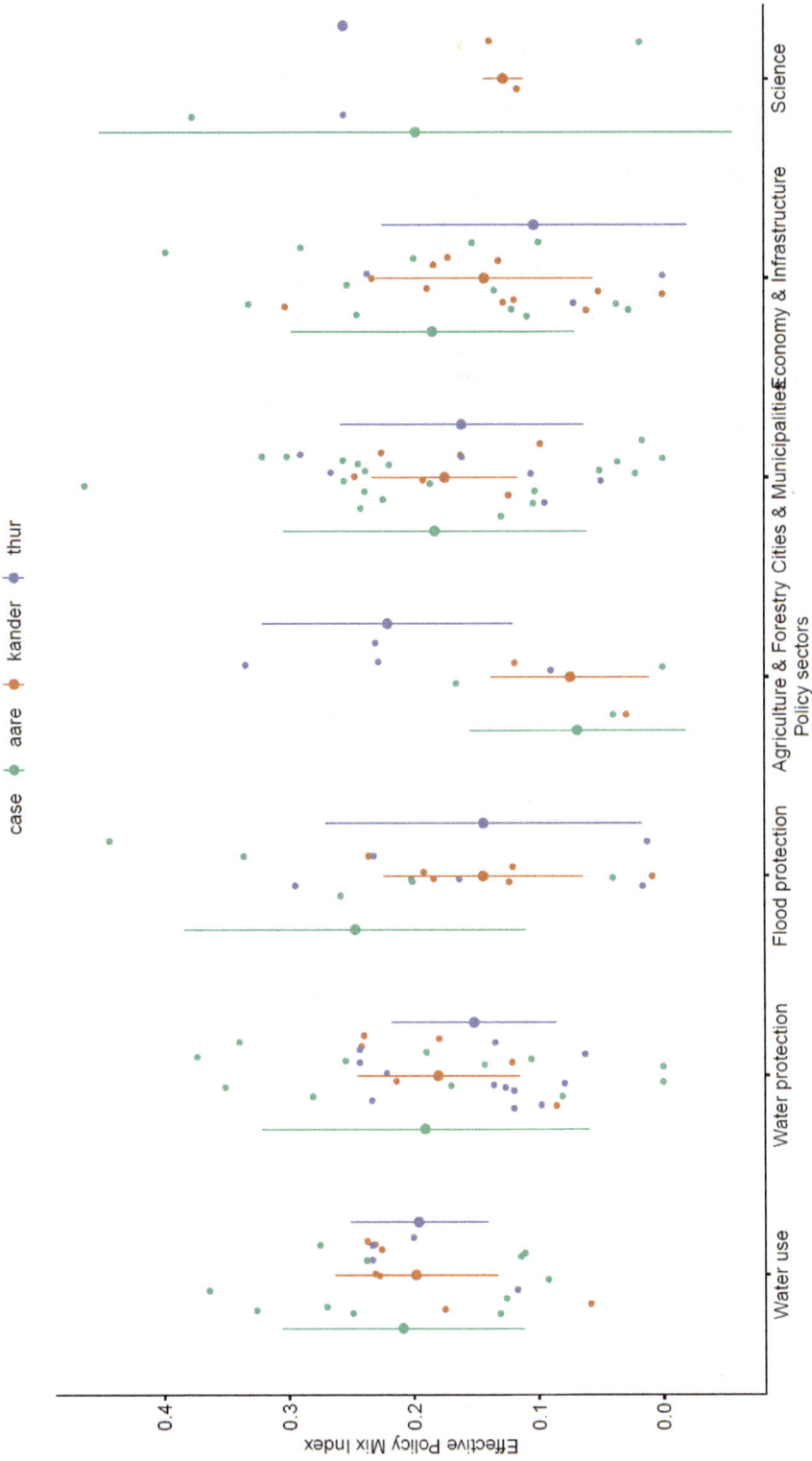

Figure 2: Actors preferences by sector and sub-catchment.

As for actors' preferences by sector, level, and sub-catchment, the interviews shed light on two points. First, actors' heterogeneous preferences in the Flood protection sector are notable, mainly the high cantonal preferences compared with the low national together with medium regional and local preferences in the Aare sub-catchment. Several actors emphasize the canton's key role (project leader) for the flood risk management project and its strong support of a participative policy-making process. The coordinated and regulated communication between canton, municipalities, and other actors helped to weigh different interests against each other and to find a feasible solution for all affected actors. It is in the canton's interest to provide coordinated flood risk management involving as many actors as possible in the policy design process. Accordingly, the cantonal Flood protection actors in the Aare sub-catchment show higher preferences for an effective instrument mix than do actors from the other sectors, levels, and sub-catchments. Second, cantonal actors' higher preferences in the agriculture and forestry sector, mainly in the Thur sub-catchment, can be explained by the controversial discussion on land use to implement flood risk management instruments. While several actors in the interviews support the use of agricultural or forested land to implement instruments by simultaneously compensating the landowners for their losses, some actors oppose this process. In the flood risk management project in the Thur sub-catchment, the canton held a key role in the form of project leader, who efficiently negotiated with forest landowners to purchase a significant parcel of land to implement flood risk management instruments by compensating landowners for their losses. This approach helped the cantonal actors to find a comprehensive policy solution including landowners' and other actors' interests. Therefore, similar to the cantonal flood protection actors in the Aare sub-catchment, the cantonal agriculture and forestry actors in the Thur sub-catchment have higher preferences for an effective policy mix than do actors from the other sectors, levels, and sub-catchments. These two examples illustrate how actors can diverge in their preferences for an effective instrument mix, depending not only on their affectedness by a problem, but also by their role in the policy design process, i.e. their responsibility to make adequate decisions in the form of instruments.

## 6 DISCUSSION

Our results reveal that, on average, most actors express similar preferences regarding an effective instrument mix, both within and across multiple sectors and levels. However, there are two sectors – flood protection and agriculture and forestry – in which actors show more heterogeneous preferences across levels, i.e. have higher or lower preferences compared to actors' average preferences in the other sectors. These findings by sectors and levels demonstrate the following: (1) Actors' preferences within the same sector can differ substantially, regardless of whether a sector is directly or indirectly affected by flood risks. Actors' preferences are neither consistently high in the directly affected flood protection sector, nor consistently low in the other indirectly affected sectors. Therefore, hypotheses 1a and 1b can neither be fully confirmed nor fully rejected. (2) The same observation is true for actors' preferences by levels. One cannot assume that actors belonging to the national and cantonal levels are only indirectly affected by flood risks, and thus generally prefer less effective mixes, while the local and regional levels are always directly affected by flood risks, and thus prefer more effective mixes. In the examples presented above, the cantonal level is often directly affected by flood risks, such that the canton plays a key role in solving problems and negotiating with other affected actors, therefore indicating overall higher preferences for an effective mix. On the other hand, regional and local actors demonstrate both comparably high and low preferences for an effective mix. Therefore, hypotheses 2a and 2b can neither be fully confirmed nor fully rejected.

In conclusion, actors belonging to a directly or indirectly flood-affected sector or level do not necessarily share preferences for an effective instrument mix. Their preferences depend on many more factors than exclusively belonging to the same defined entity. Each of the surveyed sub-catchments has its own specific actor constellations and conflicts between certain actor groups (e.g. flood protectors versus farmers). Another factor with the capacity to influence actors' preferences is their role in the policy design process. According to the literature [26], sub-national actors can strengthen the connection between national and local levels and guarantee an efficient flow of information within the three levels. This gatekeeper role can help to build a common understanding of a current problem, and thus enhance efficient task execution and enable effective policy-making. In line with Ostrom's polycentric governance approach, medium-scale governance units in the study's context are enforced with policy-making and implementation responsibilities [2].

## 7 CONCLUSION

Complex environmental problems, such as increasing flood risks, touch upon various sectors, levels, and territories simultaneously. This complexity challenges policy makers to find adequate policy solutions. Some concepts offered in the literature provide ideas on how to approach complex environmental problems with more effective policy designs supported by multiple actors with various goals, interests, and priorities. In order to discern the likelihood of designing and implementing such encompassing solutions, this study analyzes actors' preferences for effective instrument mixes in three Swiss sub-catchment areas in the case of flood risk management. Results illustrate that actors' preferences for an effective policy mix are low, and vary substantially between sectors, levels, and sub-catchments. Cantonal actors in some specific sectors show slightly higher preferences for an effective mix than the other actors do, which can mainly be attributed to case specificities and their key role in policy design processes. Taking into account such case-specific policy contexts and long-standing arrangements seems to be of major importance in evaluating actors' preferences for an effective mix. The study's findings also show that, in general, actors' preferences continue to promote sectoral and level-oriented single instruments rather than a coordinated and effective policy mix. Actors are not (yet) willing to support effective policy solutions and move towards integrated flood risk management in Switzerland.

This study's findings could have practical implications for future research on the design of effective policy mixes to address complex problems. In fact, actors' preferences to implement instruments in an effective mix are comparatively low for the case of Swiss flood risk management. Nevertheless, an analysis of the index's single indicators shows that there is potential for solving complex problems via effective instrument mixes. Adding a third indicator *balance* to the well-known combination of the indicators *density* and *intensity* proves a difficult endeavor, but as the study's results indicate, actors in Swiss flood risk management often support balanced instrument types in an instrument mix. The indicator *balance* reflects the extent to which actors are willing to compromise with other actors and accept other goals, interests, and priorities in an effective instrument mix. In a number of in-depth interviews, actors confirm this tendency by stating that several instrument types in flood risk management must be combined in order to achieve a feasible instrument mix, rather than to solely rely on the traditional technical instruments. It is evident, however, that there exists neither a one-size-fits-all design nor an ideal model to promote effective policy solutions [7].

REFERENCES
[1]   Ingold, K., Driessen, P.P.J., Runhaar, H.A.C. & Widmer, A., On the necessity of connectivity: Linking key characteristics of environmental problems with governance modes. *Journal of Environmental Planning and Management*, **62**, pp. 1821–1844, 2019. DOI: 10.1080/09640568.2018.1486700.
[2]   Ostrom, E., Polycentric systems for coping with collective action and global environmental change. *Global Environmental Change*, **20**, pp. 550–557, 2010. DOI: 10.1016/j.gloenvcha.2010.07.004.
[3]   Varone, F., Nahrath, S., Aubin, D. & Gerber, J., Functional regulatory spaces. *Policy Sciences*, **46**, pp. 311–333, 2013. DOI: 10.1007/s11077-013-9174-1.
[4]   Jenkins-Smith, H., Nohrstedt, D., Weible, C. & Sabatier, P., The advocacy coalition framework: Foundations, evolution, and ongoing research. *Theories of the Policy Process*, 3rd ed., eds P.A. Sabatier & C.M. Weible, Westview Press: New York, pp. 183–223, 2014.
[5]   Bressers, H. & O'Toole, L.J., Instrument selection and implementation in a networked context. *Designing Government: From Instruments to Governance*, eds P. Eliadis, M. Hill & M. Howlett, McGill-Queen's University Press: Montreal, pp. 132–153, 2005.
[6]   Knill, C., Schulze, K. & Tosun, J., Regulatory policy outputs and impacts: Exploring a complex relationship. *Regulation and Governance*, **6**, pp. 427–444, 2012. DOI: 10.1111/j.1748-5991.2012.01150.x.
[7]   Howlett, M., Beyond Good and evil in policy implementation: Instrument mixes, implementation styles, and second generation theories of policy instrument choice. *Policy and Society*, **23**, pp. 1–17, 2004. DOI: 10.1016/S1449-4035(04)70030-2.
[8]   Howlett, M., Mukherjee, I. & Woo, J.J., From tools to toolkits in policy design studies: The new design orientation towards policy formulation research. *Policy and Politics*, **43**, pp. 291–311, 2015. DOI: 10.1332/147084414X13992869118596.
[9]   Hood, C., *The Tools of Government*, Chatham House Publishers: Chatham, NJ, 1986.
[10]  Vedung, E., Policy instruments: Typologies and theories. *Carrots, Sticks and Sermons: Policy Instruments and Their Evaluation*, eds M.L. Bemelmans-Videc, R.C. Rist & E. Vedung, Transaction Publishers: New Brunswick, NJ, pp. 21–58, 1998.
[11]  Schneider, A. & Ingram, H., Behavioral assumptions of policy tools. *The Journal of Politics*, **52**, pp. 510–529, 1990.
[12]  Salamon, L.M., *The Tools of Government: A Guide to the New Governance*, Oxford University Press: Oxford, 2002.
[13]  Howlett, M., From the "old" to the "new" policy design: Design thinking beyond markets and collaborative governance. *Policy Sciences*, **47**, pp. 187–207, 2014. DOI: 10.1007/s11077-014-9199-0.
[14]  Bressers, H. & O'Toole, L.J., The selection of policy instruments: A network-based perspective. *Journal of Public Policy*, **18**, pp. 213–239, 1998.
[15]  Knill, C., Schulze, K. & Tosun, J., Politikwandel und seine messung in der vergleichenden staatstätigkeitsforschung: Konzeptionelle probleme und mögliche alternativen. *Politische Vierteljahresschrift*, **51**, pp. 409–432, 2010. DOI: 10.1007/s11615-010-0022-z.
[16]  Schaffrin, A., Sewerin, S. & Seubert, S., Toward a comparative measure of climate policy output. *Policy Studies Journal*, **43**, pp. 257–282, 2015. DOI: 10.1111/psj.12095.
[17]  Schmidt, T.S. & Sewerin, S., Measuring the temporal dynamics of policy mixes: An empirical analysis of renewable energy policy mixes' balance and design features in nine countries. *Research Policy*, **48**, 103557, 2019. DOI: 10.1016/j.respol.2018.03.012.

[18]    Landry, R. & Varone, F., The choice of policy instruments: Confronting the deductive and the interactive approaches. *Designing Government: From Instruments to Governance*, eds P. Eliadis, M. Hill & M. Howlett, McGill-Queen's University Press: Montreal, pp. 106–131, 2005.

[19]    Metz, F. & Ingold, K., Policy instrument selection under uncertainty: The case of micropollution regulation. *Annual Meeting of the Swiss Political Science Association*, Bern, Switzerland, 30–31 Jan. 2014.

[20]    Knoepfel, P. & Bättig, C., *Environmental Policy Analyses: Learning from the Past for the Future – 25 Years of Research*, Springer-Verlag: Berlin, Heidelberg, 2007.

[21]    Ingold, K., Fischer, M. & Cairney, P., Drivers for policy agreement in nascent subsystems: An application of the advocacy coalition framework to fracking policy in Switzerland and the UK. *Policy Studies Journal*, **45**, pp. 442–463, 2017. DOI: 10.1111/psj.12173.

[22]    Lubell, M., Feiock, R.C. & de La Cruz, E., Local institutions and the politics of urban growth. *American Journal of Political Science*, **53**, pp. 649–665, 2009. DOI: 10.1111/j.1540-5907.2009.00392.x.

[23]    Zaugg Stern, M., Philosophiewandel im schweizerischen wasserbau: Zur vollzugspraxis des nachhaltigen hochwasserschutzes. Abteilung Humangeographie, Geographisches Institut der Universität Zürich, 2006.

[24]    Jong, P. & van den Brink, M., Between tradition and innovation: Developing flood risk management plans in The Netherlands. *Journal of Flood Risk Management*, **10**, pp. 155–163, 2017. DOI: 10.1111/jfr3.12070.

[25]    Henstra, D., The tools of climate adaptation policy: Analysing instruments and instrument selection. *Climate Policy*, **16**, pp. 496–521, 2016. DOI: 10.1080/14693062.2015.1015946.

[26]    Ingold, K., How involved are they really? A comparative network analysis of the institutional drivers of local actor inclusion. *Land Use Policy*, **39**, pp. 376–387, 2014. DOI: 10.1016/j.landusepol.2014.01.013.

# SECTION 2
# FLOOD WARNING
# AND FORECASTING

# KEEPING EVERYONE ON THE SAME PAGE DURING FLASH FLOOD EMERGENCIES

BAXTER E. VIEUX & JEAN E. VIEUX
Vieux & Associates, Inc., Oklahoma, USA

## ABSTRACT
Emergency managers need to know when and where flooding is likely to occur to effectively respond to hazards. Understanding flood risk is aided by the use of a common operating picture (COP) that helps keep everyone on the same page during a flood emergency. Emergency response is aided by bringing together diverse information including distribution of rainfall and predicted flood inundation. The complexity of information sources that can benefit flood management decisions requires thoughtful integration and management. Knowing where and when a stream is likely to overtop its banks, or if road intersections are forecast to flood, makes advanced actions possible and helps protect citizens and property. Public flood warnings are generally the responsibility of national water management, or weather forecasting agencies. However, flood forecasting information services are increasingly being developed for and by municipalities, flood control districts, or private entities requiring more specificity of location, timeframe, and type of flood information. A gap analysis of a flood warning system after a major flood resulted in the development of a COP as a cloud-hosted web application. The results presented demonstrate effectiveness of the COP, which is an integrated system that supports many aspects of an early warning system used internally by agencies responsible for emergency management.
*Keywords: floods, forecasting, decision support, computer information systems, cloud-based integration, emergency management, operational dashboard, real-time data, common operating picture.*

## 1 INTRODUCTION
Occurrences of heavy rainfall and flood events are increasingly impacting people and infrastructure, especially as urban development encroaches into floodplains [1], [2]. Ahmadalipour and Moradkhani [2] analyzed flash flood occurrence from 1996–2017 and found that 99% of flash floods in the continental US (CONUS) occur due to combination of rain on burned area, rain on snow, tropical storm, or generally heavy rainfall. Texas, Oklahoma, and Arizona were identified as the top flashflood prone states in their analysis, with frequent events and abundant casualties. Climate change has affected the number of flashflood events, increasing by 1% over the 22-year period studied by Ahmadalipour and Moradkhani [2]. They found that total property damages caused by flashfloods also exhibited a significant positive trend in the past 22 years. Occurring in 2017 along the Texas Gulf Coast, Hurricane Harvey topped all other flash floods in terms of property damage within the period studied. Global recognition to reduce disasters of all causes is represented by the Sendai Framework for Disaster Risk Reduction (SFDRR) adopted by signatory states in 2015 [3]. The SFDRR establishes targets for signatory states to accomplish disaster reduction including: "Substantially increase the availability of and access to multi-hazard early warning systems and disaster risk information and assessments to the people by 2030". Thus, an early warning system (EWS) is recognized globally as an important tool for disaster reduction.

An EWS has the potential to save lives particularly if the warnings are effectively communicated to vulnerable populations [4], which is also reported by Terti et al. [5] in the US. Perhaps typical of the US, they found a high number of fatalities associated with being in a vehicle. Further, the number of fatalities depends on the time of day, nighttime flash floods produce more fatalities than in the daytime [2], [5]. These factors related to flood

WIT Transactions on The Built Environment, Vol 194, © 2020 WIT Press
www.witpress.com, ISSN 1743-3509 (on-line)
doi:10.2495/FRIAR200051

vulnerability and social science aspects emphasize the importance of the human dimension when designing flood risk communication that reaches the most vulnerable. This need is predicated on the target audience of the EWS, whether it is public facing or for internal agency use. Considering the public's perception requires adding a social science dimension to create an effective warning system [6]. Public facing systems can suffer from ineffective communications, either because the public did not take appropriate actions, did not perceive, understand or trust the warning information, or because they never received the warning in a timely manner or at all (see for example the survey results after the 2013 Front Range Flood in Colorado, US by [7]).

At the local level there are several examples of EWS setup and operations by local organizations such as municipalities for internal use in coordination of emergency response for the City of Austin, Texas flood early warning system [8]. Operational inundation mapping is becoming more important because of its value in communicating risk for emergency response as described by Greene et al. [9]. Thielen et al. [10] describe the development of the European Flood Alert System (EFAS), which aims at increasing preparedness for floods in trans-national European river basins. At the national level, flood forecasting systems have been deployed such as SCHAPI in France [11], ISOK in Poland [12], in Barcelona, Spain [13], in Germany [14], and in Western Puerto Rico US [15]. The United States National Weather Service (USNWS) has been deployed the National Water Model (NWP) for the CONUS. This ambitious project relies on hydrometeorological forecasts as input to a physics-based model WRF-Hydro [16]. Forecasts are produced for 2.7 million stream segments, with many basins that have not benefitted from model calibration, or where flood control reservoirs are only beginning to be modeled and are not yet operational [17].

Parker [4] identified barriers that may limit the effectiveness of flood warnings and appropriate warning include: (1) translating technical flood forecasts into warnings that are readily understandable by the public; (2) taking legal responsibility for warnings and their dissemination; (3) raising flood-risk awareness; (4) designing effective flood warning messages; (5) knowing how best and when to communicate warnings; and (6) addressing uncertainties surrounding flood warnings. While there have been advances in computer modelling and mapping of flood events, accurate forecasting of surface-water floods can exceed technical or modelling capabilities. Simpler approaches to an EWS can consist of a network of sensors, such as rain gauges and stream level sensors, and some that employ weather radar (see [18], [19]).

As frequently observed, uncertainty can plague hydrologic forecasting especially if physical processes are not well captured in the model structure, such as subsurface runoff. Bandaragoda et al. [20] and Beven [21] observed that hydrologic prediction of subsurface runoff processes can become intractable at the catchment scale. Two principal reasons are: (1) the unknown a-prior value and distribution of parameters controlling shallow water table or aquifer properties that are not well understood; or (2) hydrometeorological forcings, precipitation or snow melt, may not be adequately characterized over the catchment [21]. Excellent hydrologic prediction accuracy from operational distributed modelling was found by Looper and Vieux [22] in urban and peri-urban basins during Tropical Storm Hermine (2010) in Central Texas, US. Analysis of operational results that did not involve adjustment during this extreme event revealed the accuracy associated with two hydrometeorological forcings: (1) gauge adjusted radar rainfall (GARR); and (2) a gauge-only network. To gain equivalent accuracy to that obtained with GARR, the gauge network, though relatively dense, would require further densification. Further analyzes concerning physics-based distributed modelling utilizing weather radar inputs may be found in [23]–[26]. While uncertainty in hydrologic forecasts is unavoidable, incorporating uncertainty as a part of the forecast information has met with limited success. Berthet et al. [27] describe an assessment of incorporating uncertainty measures communicated as part of the forecast information but found only limited acceptance among governmental flood forecasting centers and hydroelectricity suppliers because of potential confusion associated with the uncertainty bounds communicated. Vieux [28] describe case studies involving forecast uncertainty in real-time hydrologic forecasting of reservoir inflows to hydroelectric facility with gated dam operations. The operators later declined receiving 20%, 50%, 80% uncertainty bounds in favor a single forecast representing the 50% forecast precipitation as input to the hydrologic forecast of reservoir inflows over a forecast horizon of 3–7 days.

Modern approaches to mitigating flood impacts include information technology embodied by an EWS, sometimes alone or in addition to physical flood control works such as dams, levees, detention ponds, or channel improvements. In the case where pluvial flooding is the main focus, an EWS characteristically contains (1) hydrometeorological input; (2) hydrologic and hydraulic modelling; (3) warnings communicated via text messaging, email, direct phone calls; and (4) inundation mapping that may include damage estimates [4]. One approach to reducing impacts from flooding is to minimize the public's exposure by closing flooded intersections, and warning stakeholders in the affected basins. Emergency responders benefit from knowing when and where flooding is likely to occur, so that effective flood response services can be delivered. Increasingly, an EWS is used to deliver critical flood information in real-time via a dashboard containing essential flood and rainfall-runoff information (additional details in [8]).

Recent EWS technological innovations leverage cloud-based integration of model-based predictive stream flow, rain gauge and radar rainfall, computer-aided dispatch, tweets, high-resolution physics-based flood forecast flow and stage, and predictive inundation mapping [1]. Detailed flood inundation mapping helps focus emergency response to areas where a flood is expected, helping to communicate risk within internal agency operations. The following sections provide an overview of the common operating picture (COP) concept, and description of key elements implemented to provide a cloud-based system that aids in decision-making during flood emergencies.

## 2  METHODOLOGY

A gap analysis was performed to develop design objectives of a COP for use by a municipality for internal agency coordination and operations management. The use case is to provide decision support for where to position barricade crews, make highwater rescues, and prepositioning of assets. The following objectives and solutions provide a systematic approach taken in its design with resulting web-application pages shown that are now in operational use. Table 1 summarizes gaps identified that were identified by the agency regarding an existing flood management system, and solutions identified for improvement and implementation as a COP flood management system. The existing system was operational during a major flood in 2013 in Central Texas USA, when the municipality experienced need for improving technology to aid their emergency operations. For each Gap (1–4) identified, a corresponding Solution (1–4) is presented in the sections that follow.

### 2.1  Display timeframe

To satisfy Gap 1, the solution past, present and future timeframes are useful so that replay can show storm and flood evolution leading to the present or current conditions. Future precipitation forecasts when added to current rainfall extends available lead-time before flooding reaches or exceeds threshold values of depth and flow as shown in Fig. 1.

Table 1:  Gap analysis identifying needs and solutions for development of the COP.

| Objective | Solution |
|---|---|
| **Gap 1:** The current system did not support replay or easily support knowing where and when flooding was occurring. It was not integrated causing staff to not have access to crucial information (e.g. geo-located flood reports) or must rely on multiple computer system display screens. | • Solution 1: Display timeframe. The system needs to display:<br>• Forecast (maximum) flood depth<br>• Current inundation and road closures<br>• Future locations of flooded roads and streams |
| **Gap 2:** Needed data sources which were not integrated or available during emergency operations centre activation. | **Solution 2:** Diverse data source integration. Integration of tweets and 311 calls from multiple computer systems. |
| **Gap 3:** Corporate knowledge was not integrated into the system such as how upstream flood depths could affect downstream out-of-bank flooding, road closures, or structure flooding. | **Solution 3:** Rules engine. Ability to write rules that affect system operation:<br>• User can add and modify rules based on modelled/observed flood levels<br>• "What-if" functionality<br>• Automated alert notifications |
| **Gap 4:** After-action reporting was not supported by the system for upper management briefing, storm report, or flood damage assessment. | **Solution 4:** Automated reports. After-action reports will provide summaries of rainfall storm totals, structures flooded, road closures, low water crossing device activation, and inundation polygons. |

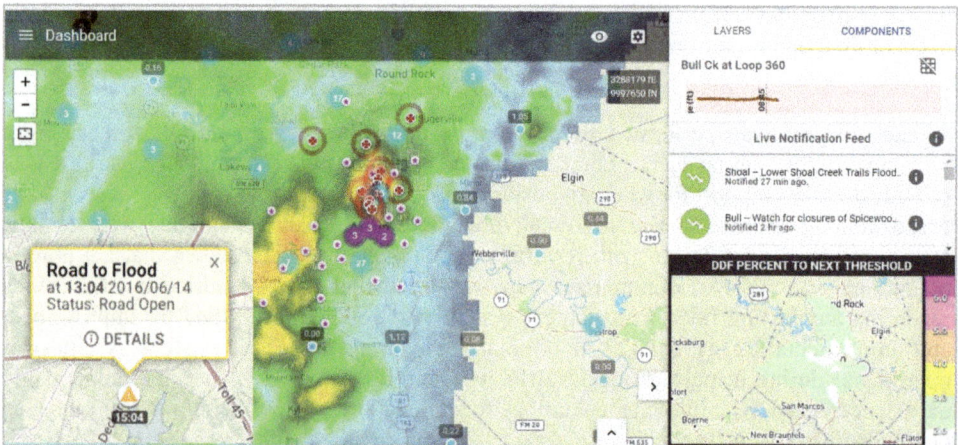

Figure 1:    Real-time dashboard with map display of current radar rainfall (map background) with tweets/alerts indicated by icons in map (red cross with pulsating circles), road to flood (bottom left) countdown timer with 13:04 minutes left, forecast and observed hydrographs (top right) with alert log shown with green circles and arrows indicating stage direction (up/down), bottom right shows map of depth duration frequency (light green watersheds < 2 years).

## 2.2  Diverse data source integration

There is a diversity of data sources that can be useful during a flood emergency but require integration from other computer systems. Fig. 2 presents a conceptualization of tweets and citizens flood "complaints" (referred to as 311 calls) (left panel), map display of icons and flooding (centre), and a "feed" of notifications as they are produced last-on-top (right).

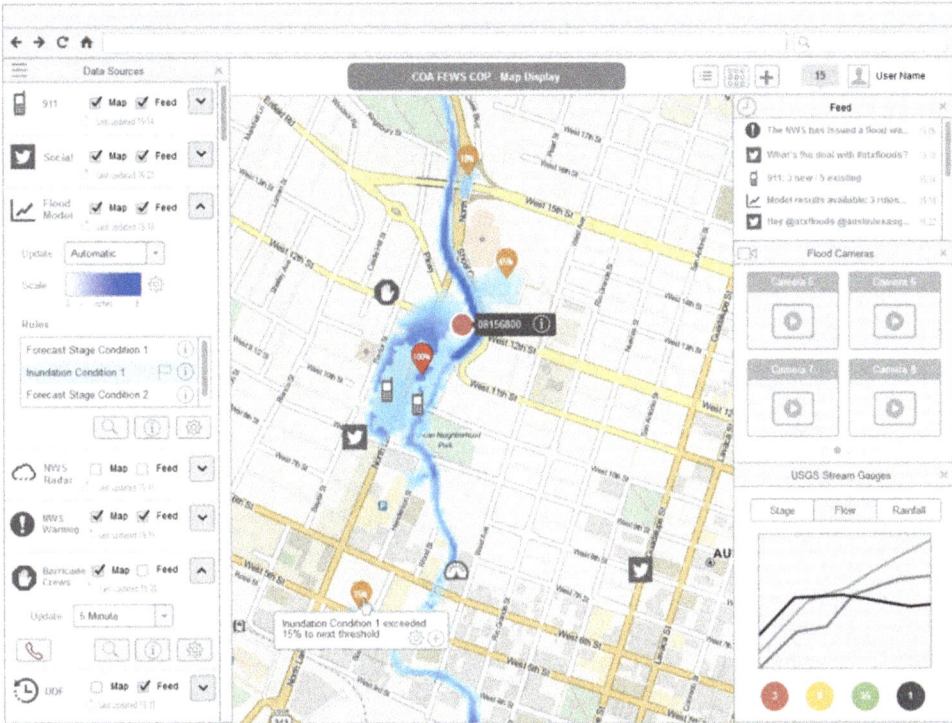

Figure 2:    Map view showing integration of data elements (Barricades/311 calls/Tweets/ Streamgauge) and display configuration.

## 2.3  Rules manager

This element provides the user with the ability to write and save rules that affect system operation. For example, the user could add or modify rules based on modelled/observed flood levels, causing an automated alert notification to appear as a flashing map icon or as a notification that a threshold has been exceeded. Fig. 3 shows the control configuration used to set the Data Source (observed or simulated stage), Location (stream gauge), Rule Type (stage threshold) and its value, together with Notifications desired such as "Flag on Map", "Show in Feed", "Send Email/MMS", and the type of "Map Icon".

## 2.4  Automated reports

Automated reporting supports generation of reports after a flood emergency. Both spatial and tabular information contained in the database is extracted and formatted as an editable

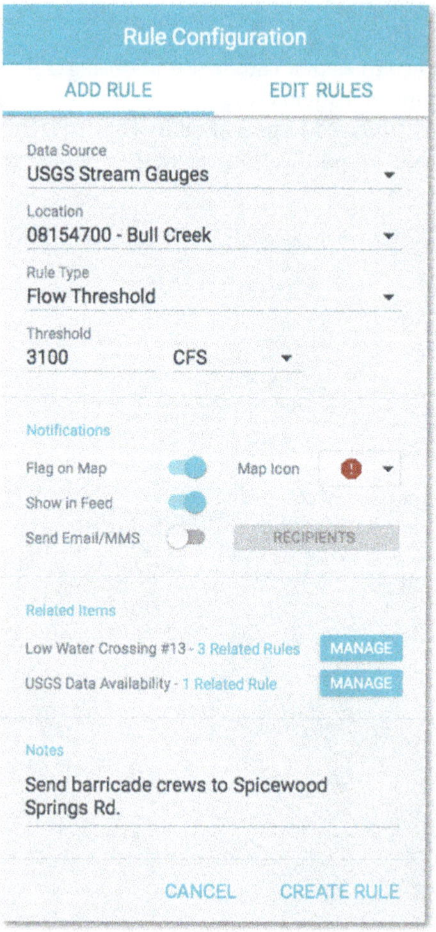

Figure 3:    Rules engine allows configuration: Adding/editing/writing rules for system operation. Data source, threshold, and notifications are shown with notes on where to send barricade crews.

document. Report content includes tabular data summaries that include road closures, rules triggered, and structure inundation with preliminary damage estimates. Fig. 4 shows a rainfall total based on weather radar and rain gauge measurement, accumulating 500 mm over several days. The gridded maps were produced in real-time with rain gauge measurements to correct bias (filled circles), called gauge-adjusted radar rainfall.

## 2.5  Cloud-hosting

A key system characteristic of the COP is integration of multiple data sources. Because integration was identified as a gap, cloud-hosting became necessary and advantageous solution. This design makes possible the integration of real-time predictive flood information data sources, while maintaining user access during periods of high usage, such as flood emergencies.

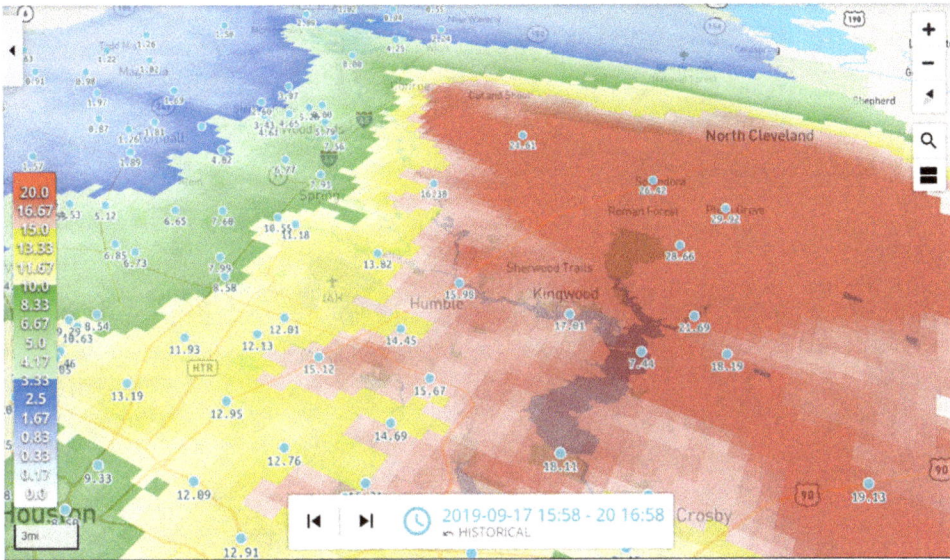

Figure 4:  Automated report of a storm total greater than 500 mm (1 inch = 25.4 mm).

## 3  RESULTS

Illustrative examples of system performance are presented as follows from the COP and several other EWS's that rely on bespoke radar rainfall systems as input to a physics-based distributed model, V*flo*® [8] that is operational for multiple organizations. Fig. 5: shows two hydrographs for Brays Bayou (256 km$^2$ drainage area) at Main Street (bottom) and upstream at Gessner (top) with a 135.9 km$^2$ drainage area. The operational results are not adjusted in real-time and. The model shows excellent agreement (observed shown as dark red points) from its starting stage on 25 August to 27 August 2017 during beginning portions of Hurricane Harvey which lasted nearly 5 days over Houston Texas. The vertical line is the "now line" at 21:45 CDT on 26 August indicates the time of current rainfall. It is interesting to note that these two stream gauges peak at nearly the same time due to the shape of the watershed. The forecast hydrograph (dark blue) is coincident with observed. At Gessner, the forecast hydrograph is aligned with initial rise in observed. The observed stops short of the "now line" with the last observation, lagging about 1 hour. Lead-time is demonstrated by observing that the forecast peak just under flood stage at Gessner and above flood stage (warning with red shading and watch with yellow). Emergency action and awareness is achieved almost an hour before the flood arrives. Lead-time is closely tied with the location of intense rainfall in a watershed relative to the forecast stream gauge or location. In the case of Harvey, an intense band of rainfall was traveling downstream in the watershed, and with intensification just above the Main Street location, little hydraulic lead-time (less than 1 hour) was achieved. Consequently, hydrologic forecasts based on current rainfall (dark blue) and forecast rainfall (yellow) show little departure after the "now-line" until several hours later as evidenced by the second peak that follows after the first.

Another municipal system, located in Central Texas, also organizes its flood warning operations around an EWS composed of GARR and forecast rainfall from numerical weather prediction and nowcasting [8]. Operational flood emergency management and response is coordinated with the aid of a COP described above in the gap analysis. From forecast

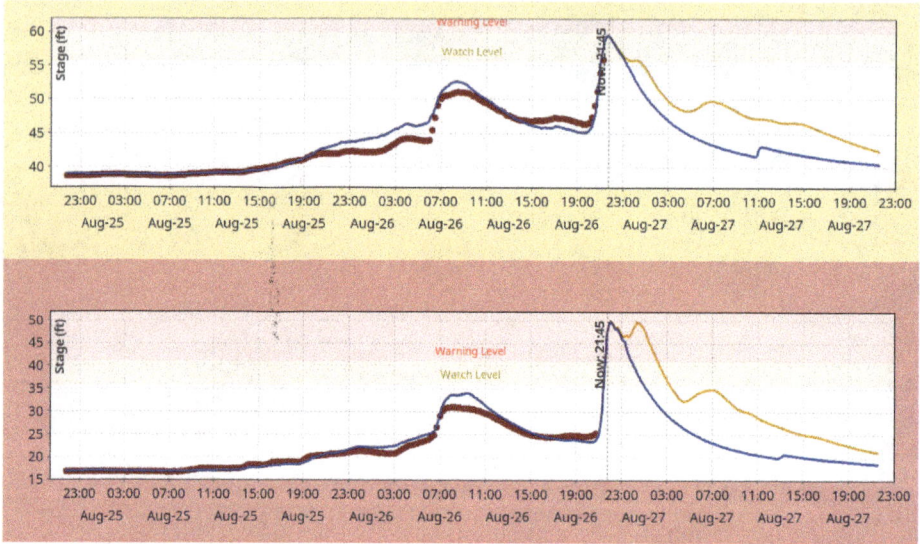

Figure 5:   Operational forecast model results showing agreement between observed hydrographs (dark red points) and forecast hydrographs based on current and forecast rainfall (yellow) 25–27 August 2017 during Hurricane Harvey in Houston, Texas (1 ft = 0.3048 m referenced to mean sea level).

locations are entered into the Rules Manager based on modelled or observed hydrograph upstream to alert staff of anticipated downstream impacts. For example, at an upstream location, if forecast and/or observed stage exceeds a given threshold, then downstream street intersections are barricaded. This alert notification captures user experiences and integrates it within the system for future use, thus capturing institutional knowledge. There are 166 such rules entered into the system by operational flood staff to assist operations during flood emergencies. Fig. 6 shows the results of Hurricane Harvey after it travelled inland from the Gulf Coast. Its strength was considerably diminished but still produced flooding above flood stage watch and warning levels at 4 ft (1.2 m) and 5 ft (1.5 m), respectively over a prolonged period of several days in an urbanizing watershed comprising only 5.6 km$^2$ drainage area. Rainfall accuracy and relatively impervious soils modelled with hydraulics derived from 5 m resolution gridded elevation data produce exceptionally accurate forecast stage, again without a forecaster in the loop adjusting the model during or after operations. Results shown are "as-is" obtained during operations during Hurricane Harvey. The dark red points indicate observed stage while the green hydrograph is simulated with a USNWS weather radar (NEXRAD) product called DPR which relies on radar with dual polarization to estimate rainfall intensities [19], [28]. This product is independent of the local organization's rain gauge network, while the blue hydrograph is produced with a GARR product that is customized for the client and created operationally with multi-agency rain gauge networks integrated with USNWS NEXRAD to produce an operational near-real-time (NRT) bias-corrected hydrometeorological forcing for distributed model input. The light blue streaks are the hydrographs progressively generated with rainfall known at the time. The net result of these intermediate hydrographs when viewed for the totality of the storm duration, is a progression of hydrographs that when combined, produce the final hydrograph result (dark blue).

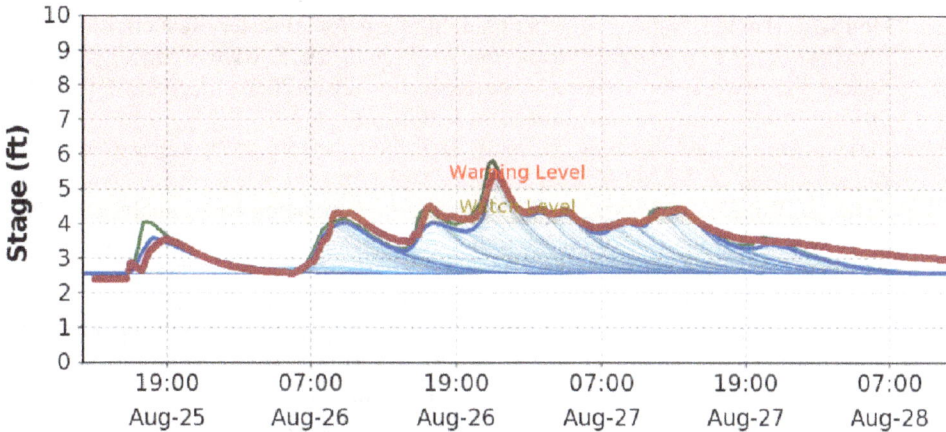

Figure 6:  Results of Hurricane Harvey after travelling inland from the Gulf Coast with diminished strength but still produced flooding above flood stage during Harvey (1 ft = 0.3048 m referenced to local channel bottom).

## 4 SUMMARY

Managing flood emergencies with the aid of an EWS is becoming more widely adopted. Because heavy rainfall and flooding are affecting more people and infrastructure, organizations are turning to an EWS such as the COP described herein to help anticipate, effectively respond, and manage post-flood recovery. From a gap analysis, solutions were identified for development of the COP that delivers predictive flood information in near real-time to operational personnel. To accomplish objectives, NRT radar and rain gauge inputs are used for predictive modelling of depth and flow hydrographs at many locations through urban and peri-urban watersheds affecting a flooding. From modelled and observed data, predictive inundation maps and alert notifications are generated and mapped. This provides context and situational awareness during flood emergencies. Rules written by the end-user help capture experience and institutional knowledge, thus improving system effectiveness. After-Action reporting is facilitated through automated summaries of rainfall total images and values, road closures, rules triggered during the event, and tabulation of preliminary flood damages from flood levels estimated above finished floor elevations. Integration of data and information from multiple computer systems, cloud-hosting of the system provides high availability to agency personnel within a single interface. The COP solution has been operational since 2016 and continues to help everyone keep on the same page during flood emergencies.

## REFERENCES

[1]  Vieux, B.E., Smith, R.L. & Vieux, J.E., When water becomes a hazard: A common operating picture helps keep everyone on the same page. *92nd Annual Water Environment Federation's Technical Exhibition and Conference, WEFTEC*, 2019.

[2]  Ahmadalipour, A. & Moradkhani, H., A data-driven analysis of flash flood hazard, fatalities, and damages over the CONUS during 1996–2017. *Journal of Hydrology*, **578**, p. 124106, 2019.

[3]   UNISDR (United Nations International Strategy for Disaster Reduction), *Sendai Framework for Disaster Risk Reduction 2015–2030*, 2015. www.wcdrr.org/uploads/Sendai_Framework_for_Disaster_Risk_Reduction_2015-2030.pdf. Accessed 11 May 2015.

[4]   Parker, D.J., Flood warning systems and their performance. *Oxford Research Encyclopaedia of Natural Hazard Science*, 2017.

[5]   Terti, G., Ruin, I., Anquetin, S. & Gourley, J.J., A situation-based analysis of flash flood fatalities in the United States. *Bulletin of the American Meteorological Society*, **98**(2), pp. 333–345, 2017.

[6]   Bodoque, J.M. et al., Improvement of resilience of urban areas by integrating social perception in flash-flood risk management. *Journal of Hydrology*, **541**, pp. 665–676, 2016.

[7]   Morss, R.E., Mulder, K.J., Lazo, J.K. & Demuth, J.L., How do people perceive, understand, and anticipate responding to flash flood risks and warnings? Results from a public survey in Boulder, Colorado, USA. *Journal of Hydrology*, **541**, pp. 649–664, 2016.

[8]   Vieux, B.E., Janek, S. & Vieux, J.E., How one community uses radar hydrology to cope with flooding: The city of Austin's flood early warning system. *27th Conference on Hydrology*, Annual Meeting of the American Meteorological Society, 2013.

[9]   Greene, P.S. et al., Urban flooding and climate change: Visualizing the impacts. Research Report, 2015. http://eos.ou.edu/hazards/urbanflooding/.

[10]  Thielen, J., Bartholmes, J., Ramos, M.-H. & de Roo, A., The European flood alert system: Part 1 – Concept and development. *Hydrology and Earth System Sciences Discussions, European Geosciences Union*, **5**(1), pp. 257–287, 2008.

[11]  Corral, C., Berenguer, M., Sempere-Torres, D., Poletti, L., Silvestro, F. & Rebora, N., Comparison of two early warning systems for regional flash flood hazard forecasting. *Journal of Hydrology*, **572**, pp. 603–619, 2019.

[12]  Goniewicz, K. & Burkle, F.M., Disaster early warning systems: The potential role and limitations of emerging text and data messaging mitigation capabilities. *Disaster Medicine and Public Health Preparedness*, **13**(4), pp. 709–712, 2019.

[13]  Llort, X., Sánchez-Diezma, R., Rodríguez, A., Sancho, D., Berenguer, M. & Sempere-Torres, D., FloodAlert: A simplified radar-based EWS for urban flood warning. *Hydroinformatics HIC*, New York, 17–21 Aug. 2014.

[14]  Hofmann, J. & Schüttrumpf, H., Risk-based early warning system for pluvial flash floods: Approaches and foundations. *Geosciences*, **9**(3), p. 127, 2019.

[15]  Molina, L.E.T., Floods forecast in the Caribbean. *Flood Risk Management*, p. 55, 2017.

[16]  Cosgrove, B. et al., An overview of the National Weather Service national water model. *AGUFM*, pp. H42B-05, 2016.

[17]  Kim, J., Read, L., Johnson, L.E., Gochis, D., Cifelli, R. & Han, H., An experiment on reservoir representation schemes to improve hydrologic prediction: coupling the National Water Model with the HEC-ResSim. *Hydrological Sciences Journal*, pp. 1–15, 2020.

[18]  Acosta-Coll, M., Ballester-Merelo, F., Martinez-Peiró, M. & la Hoz-Franco, D., Real-time early warning system design for pluvial flash floods: A review. *Sensors*, **18**(7), p. 2255, 2018.

[19]  Bedient, P.B., Huber, W.C. & Vieux, B.E., *Hydrology and Floodplain Analysis*, 6th ed., Prentice-Hall: Saddle River, NJ, 2018.

[20] Bandaragoda, C., Tarboton, D.G. & Woods, R., Application of TOPNET in the distributed model intercomparison project. *Journal of Hydrology*, **298**(1–4), pp. 178–201, 2004.

[21] Beven, K., 14 Distributed models and uncertainty in flood risk management. *Flood Risk Science and Management*, p. 291, 2011.

[22] Looper, J.P. & Vieux, B.E., An assessment of distributed flash flood forecasting accuracy using radar and rain gauge input for a physics-based distributed hydrologic model. *Journal of Hydrology*, **412**, pp. 114–132, 2012.

[23] Vieux, B.E., Imgarten, J.M., Looper, J.P. & Bedient, P.B., Radar-based flood forecasting: Quantifying hydrologic prediction uncertainty in urban-scale catchments for CASA radar deployment. *World Environmental and Water Resources Congress 2008*, Ahupua'A, pp. 1–10, 2008.

[24] Looper, J., Vieux, B. & Moreno, M.A., Evaluating precipitation uncertainties using the Vflo hydrologic model. 23rd *Conference on Hydrology, American Meteorological Society 2009 Annual Meeting*, ASCE, Reston, VA, Jan. 2009.

[25] Vieux, B. & Vieux, J., In pursuit of reliable flood prediction. *Flood Risk Management and Response*, p. 220, 2016.

[26] Worthington, T.A., Brewer, S.K., Vieux, B. & Kennen, J., The accuracy of ecological flow metrics derived using a physics-based distributed rainfall-runoff model in the Great Plains, USA. *Ecohydrology*, **12**(5), 2019.

[27] Berthet, L., Piotte, O., Gaume, É., Marty, R. & Ardilouze, C., Operational forecast uncertainty assessment for better information to stakeholders and crisis managers. *E3S Web of Conferences*, **7**, p. 18005, 2016.

[28] Vieux, B.E., Case studies in distributed hydrology. *Distributed Hydrologic Modelling Using GIS*, Springer: Dordrecht, pp. 211–234, 2016.

# FLOOD FORECAST IN MANAUS, AMAZONAS, BRAZIL

JUSSARA SOCORRO CURY MACIEL[1,2], LUNA GRIPP SIMÕES ALVES[1],
BRUNO GABRIEL DOS SANTOS CORRÊA[2], IRAÚNA MAICONÃ RODRIGUES DE CARVALHO[2]
& MARCO ANTÔNIO OLIVEIRA[1]
[1]Geological Survey of Brazil, Brazil
[2]Federal Center for Technological Education of Amazonas, Brazil

## ABSTRACT
The Amazon region has rivers of great contribution to the biodiversity balance and regional landscape. Among them, the stand out are the Negro River – the most extensive, the Amazon River – with the largest volume, the Purus River – a fish supplier, and the Madeira River – the most modified. Flooding in the Amazon is often widespread and sometimes severe for those who live very close to rivers, yet urban areas are often more socially impacted than rural areas, as flooding is part of the riverside population's daily life. Flood forecasting models are important for composing extreme event alerts as well as for knowledge of decision-makers, public agency representatives, and affected communities. This paper aims to present what is being done in Manaus about the Negro River floods, with the development of a flood alert dissemination project, which uses a linear regression statistical model, based on the historical series over 100 years. The follow-up to the one-year flood forecast begins the previous year, more precisely when the ebb ends and the rainy season begins. Several factors contribute to the flood event, but the main one is the rainy season and how the basin will behave during the six months of rising waters. Manaus station receives contributions from two major rivers: Negro and Solimões, which form the Amazon River downstream. The position of the city, allied to the difference in depth, speed, and flow of the rivers, added to the rainfall period in the basin reveal the flooding characteristic in this locality. Considering the last 15 years, the near-real flood forecast interval in 87% of cases reveals that the project has fulfilled the purpose of alerting the population and public representatives about the flood of Negro River.
*Keywords: extreme events, flood forecast, Amazon rivers.*

## 1 INTRODUCTION
The Amazon basin is the largest drainage basin in the world, covering about 40% of South America. The basin covers a total area of $7.05 \times 10^6$ km$^2$, approximately. The Amazon River system plays a significant role in the global hydrological cycle since its total river flow is greater than the combined flow of the ten next largest rivers. Amazon flow accounts for approximately one-fifth of the world's total river discharge to the oceans [1]. The Amazon region annual river regime is well defined and represents the variability of headwater rainfall [2]. Rainfall ranges from 1,500 to 3,000 mm annually, averaging about 2,000 mm in central Amazon [3]. Manaus, located in the Negro River, also has a so-called Equatorial regime and also presents its flood in the middle of the calendar year (June/July). The contributions of the Rio Negro are concentrated in watersheds whose main river gutters come from the northern hemisphere [2].

For the annual precipitation evolution, Sternberg [4] considered that in September is when there is the lowest rainfall, in the hydrographic area whose flow is most directly reflected in the Manaus post. In October, the basins' contribution to the south of Negro River, which receives the regime's precipitation, which grows from November to March, is accentuated. During this period there are large rainfall indices above the waterways south of Negro River, delimited by monthly isoietas of 250, 300 and up to 350 mm.

According to Witkoski [5], there are three types of rivers in the Amazon Basin: (i) the white-water rivers, which are born in the Andes and carry high fertility sediments – occur with the Japurá, Juruá, Purus, Solimões, Madeira, Amazonas and others; (ii) the black water rivers come from areas called podzolic soils, as is the case of the Negro and Urubu Rivers; and (iii) the clear water rivers that drain areas of the Central Highlands of Brazil and the Highlands of the Guianas [5].

Hydrological drought and flood extreme events were quantitatively defined to occur when daily water levels in Manaus fall below 15.8 m or rise above 29.0 m, respectively [6]. For areas of the Solimões-Amazon River in the near the municipality of Manaus, the hydrological periods were identified, as follows: (i) Flood: rising river level, between the 20 and 26 m; (ii) Flood: quota equal to or greater than 26 m; (iii) Downstream: downstream river level, between 26 and 20 m; and (iv) Drought: quota of 20 m or less [7].

In the Solimões-Amazonas River, however, there is a year-to-year variability inflow, partly related to fluctuations in rainfall. Rainfall decreases in the Amazon are partially associated with the phenomenon popularly known as "El Niño": "El Niño" seems to produce severe drought or ebb and "La Niña" cause flooding intense [7], [8].

It is known that despite differences in catchment size and timing of the flood pulse, water levels at the Manaus station on the Negro River measure fluctuations of the Solimões–Amazon main stem due to the so-called backwater effect [6]. The water levels of the lower Rio Negro are controlled by the Solimões River. During the flood period of the Solimões River, the Negro River is dammed causing flooding in the city of Manaus. The Solimões water level signal dominates the level of Negro River stations up to 400 km away [9], [10].

The Amazonian plain contains two orders of entirely different landscapes: the floodplains and the firm lands. Floodplains are wetlands in the immediate vicinity of rivers and firm lands are the highest terrain, a few meters above water, at other points they reach moderate altitudes. The floodplain rivers flow from sedimentary formations that they have deposited themselves. Some of these rivers have curve courses, such as Juruá and Purus. In the Solimões–Amazonas route, the waters are split and linked by channels, forming islands that stretch in the current direction and divide the great river [4]. The large floodplain areas along the Solimões River gutter and the low slope downstream of the basin are responsible for flood wave attenuation [2].

Droughts and floods, part of the natural climate variability in those regions, have occurred in the past and will continue to happen in the future. The inhabitants of the region are well adapted to this hydrological inter annual dynamics and, over time, have been able to develop their livelihood strategies. During the floods of 2009 and 2012, the rising levels of the Solimões and the Negro Rivers, the two main branches of the Amazon River, caused floods in urban and rural areas along the riverbanks in Peruvian, Colombian and Bolivian Amazonia [11].

In this article we present an important project developed by Geological Survey of Brazil in Manaus, that presents every year three predictions of Negro River at Manaus station, called Negro River Flood Warning. The first forecast occur in March (for example, this year happened on 29 March), the second forecast occur in Abril (e.g. on 30 April 2019) and the third in May (e.g. on 31 May 2019). The main objective of this project is to offer to the society the information about the eventual flood and a high river preview in the flood time.

## 2  NEGRO RIVER FLOOD FORECAST PROJECT

In the Amazon, the river is a means of transport, provides food and clean water. Therefore, knowing the hydrological regime is important to provide information to those who live with the flood and ebb. Negro River in Manaus spends on average 234 days of rising and 130 days of descent (Fig. 1). The climb is smooth and striking, which facilitates predictability. Since

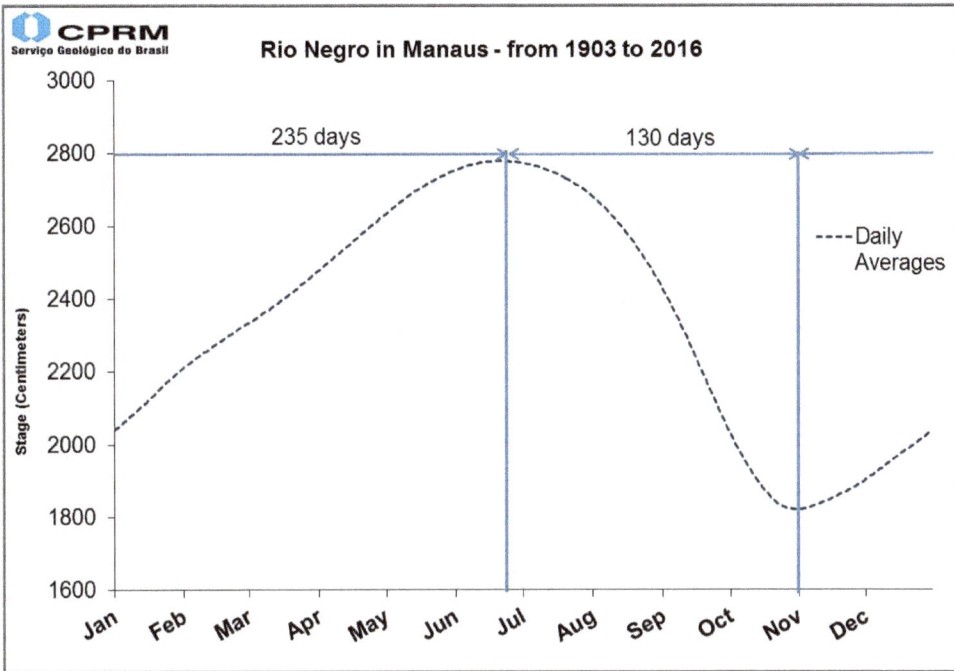

Figure 1: Length of the rise and fall of the Negro River at Manaus station.

1989, the Geological Survey of Brazil in Manaus has been developing the "Amazon Hydrological Alert System" where the annual flood and ebb monitoring process is performed in the Solimões–Negro–Amazonas system. Among the products generated by the project is the Manaus Flood Alert, which presents the forecast of the maximum quota to be reached by the Negro River in Manaus each year. The results are released to the relevant agencies and the press at the end of March, April and May, preceding the maximum Negro River quota, which usually occurs between June and July.

Thus, it is possible to do a good statistic job because, in Manaus Harbour, there is a measuring ruler, installed by the English since 1902, generating a historical series of 117 years. As Marengo et al. [11] observed the droughts of 2005 and 2010 and the floods in 2009 and 2012 were detected as the lowest daily minimum and the largest daily maximum, respectively, of the levels of the Rio Negro at Manaus, as we can see in Fig. 2. Barichivich et al. [6] examine the characteristics of individual flood events that indicates the recent floods not only occur more often but also have become more severe, such as the return period of a flood event exceeding either a duration of 70 days or a water level of 29.7 m besides they have occurred within an interval of only 3 years.

In Table 1, we present the ten floods with the highest quotas, six of which occurred in this decade, which may indicate a change in the flood pattern in Negro River. Any of the ten largest historical floods in Manaus already impacts the riverside population, as the descent is not immediate either; the physical result of the flood persists in the first month of ebb.

In 1953, Manaus and Negro River had an extreme flood recorded in 29.69 m, for more than 50 years, it was the higher level. Then in 2009, we had a great flood that exceeded 1953 and registered 29.77 m, in this event it was possible to notice that the Solimões had a different

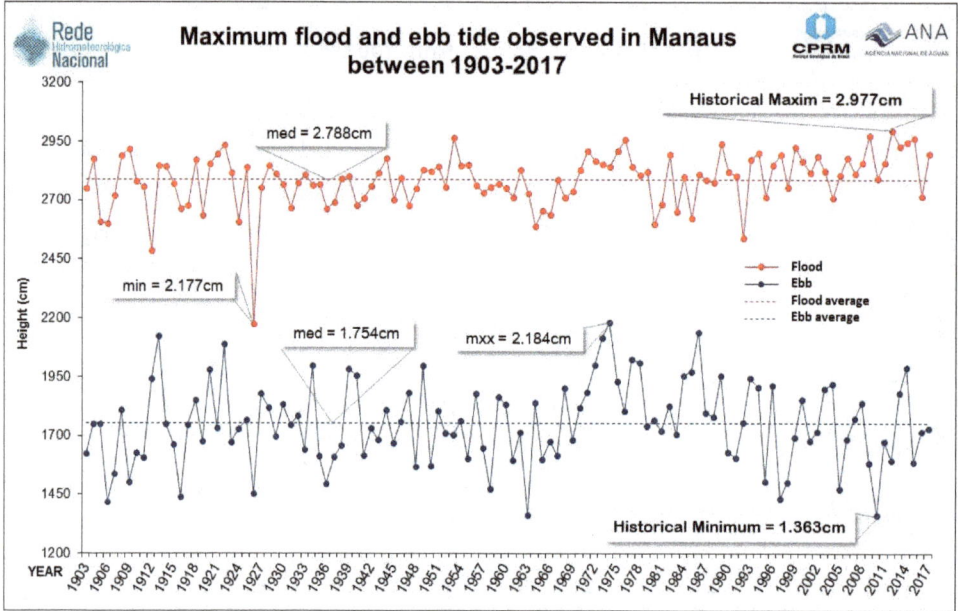

Figure 2:  Manaus Harbour maximum and minimum ruler heights.

Table 1:  The biggest floods in Manaus Station.

| Position | Year | Water level | Date | Flood days |
|----------|------|-------------|------|------------|
| 1 | 2012 | 29.97 | 29 May | 230 |
| 2 | 2009 | 29.77 | 1 July | 244 |
| 3 | 1953 | 29.69 | 9 June | 221 |
| 4 | 2015 | 29.66 | 29 June | 237 |
| 5 | 1976 | 29.61 | 4 June | 197 |
| 6 | 2014 | 29.50 | 3 July | 246 |
| 7 | 1989 | 29.42 | 3 July | 261 |
| 8 | 2019 | 29.42 | 22 June | 225 |
| 9 | 1922 | 29.35 | 17 June | 227 |
| 10 | 2013 | 29.33 | 14 June | 226 |

period of rise of the Negro, at the time when the Solimões had high quotas for the period in Manaus, the contribution of the Negro arrived and made the levels even higher this season and longer. Already in 2012, the whole basin had a great rise from February until May of the same year, when it reached a historical maximum of 29.97 m. In 2019, the first flood wave on Negro River in Manaus was influenced by Solimões and rainfall throughout the basin, similarly in the second phase of the flood (February and mid-March), however, in the third phase (April and May), what influenced the flood was the rise of the Upper Negro River in March, affected by the rains from the north of the state of Amazonas, when the waters were lowered to Manaus station, the quotas were already high, a fact that significantly influenced the mark of this flood, with a maximum record of 29.42 m, reaching 7th position in Manaus historical records.

To set up the monitoring of Negro River in Manaus, it is necessary to follow some stations located before and in different gutters that contribute to Manaus station, as illustrated in Fig. 3. These stations have rulers installed on the banks and are accompanied by local people daily. We call then as observers, through the National Water Network project supported by the National Water Agency, it is possible to pay financial aid to these observers and visit these stations at least four times a year. In the case of monitoring stations, the data is collected daily and recorded in specific logbooks. The record of rivers rise and fall in these seasons are recorded in a weekly newsletter and published on the homepage (www.cprm.gov.br) of the monitoring responsible sector and it has a great importance to the resident population, the State and city halls, as well as for academic researchers.

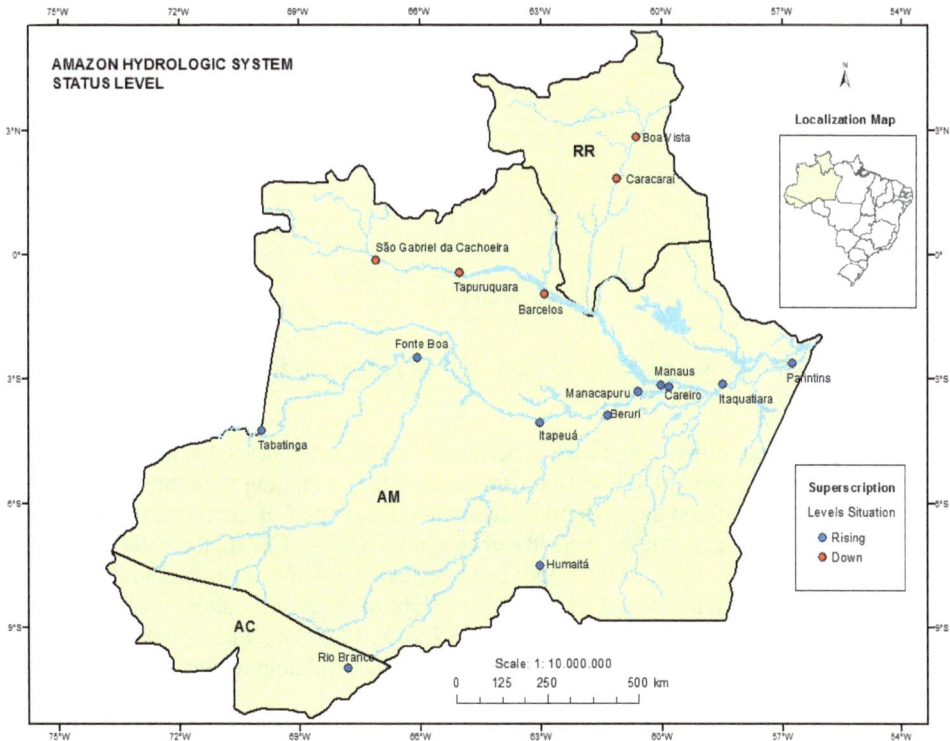

Figure 3:    Hydrological monitoring stations accomplished by flood alert. *(Source: Adapted from CPRM/SGB [12].)*

Geological Survey of Brazil forecasts are based on a simple linear regression, where the maximum annual quotas are correlated with the forecast dates, which are 31 March, 30 April and 31 May (see Fig. 4). Statistical models of simple or multiple linear regressions are efficient for long-term prediction of Manaus flood peaks, with the advantage of use facility, despite the disadvantage of not accurately determining the date of occurrence of the event.

3 HYDROLOGICAL DATA ANALYSIS OF MANAUS FLOODS
Flood frequency analysis has a local basin perspective and is based on the assumption that floods are the consequence of a random process such that are independent and identically

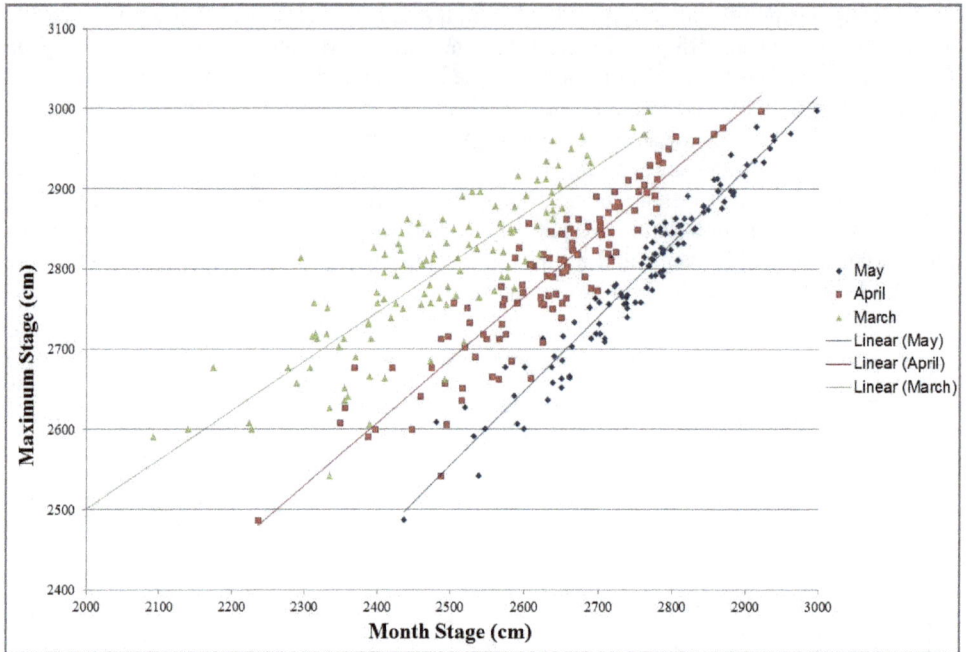

Figure 4:  Simple linear regression used in Manaus flood forecast.

distributed. However, this assumption has been questioned on the basis that climate, land use and other factors that determine flood occurrence are likely changing with time [13].

We understand the flood as a natural phenomenon that is part of the rivers dynamics, for the Manaus Civil Defence, when Negro River reaches a height of 27 m, the river population already feels the direct impact of the flood; by historical series, this quota has a return period of 10 years (T = 10). When the river reaches a height of 29 m, it is already considered an emergency water level, with a return period of 17 years (T = 17); when the river is at this stage, the sectors responsible for the city install walkways and help to smooth out the flood impact.

In Manaus, 117 floods were catalogued, of which 6% occurred in May, 75% in June and 19% in July (Fig. 5). In Table 2, we present the monthly river discharges in two stations (Paricatuba and Manacapuru) before Manaus, where we can see the increased flow of the Black and Solimões Rivers; the months of March, April and May reveal the contribution that arrives in the basin and that leads to the flood period of the year.

Considering Table 2, it is possible to see the differences in discharge between the Negro and Solimões Rivers and also the increasing flow in the years of the biggest floods (2009 and 2012). CPRM (Geological Survey of Brazil) measured these flows to understand if much difference occurs between the months preceding the flood, thus we can detach:

- In the flood of 2009, in the month of January, the discharge was the average of other years. Already in 2012, the flow in January was already with great measurements and this for both gutters (Negro and Solimões). This proves that the 2012 flood has already started with high quotas due to the basins contributions.

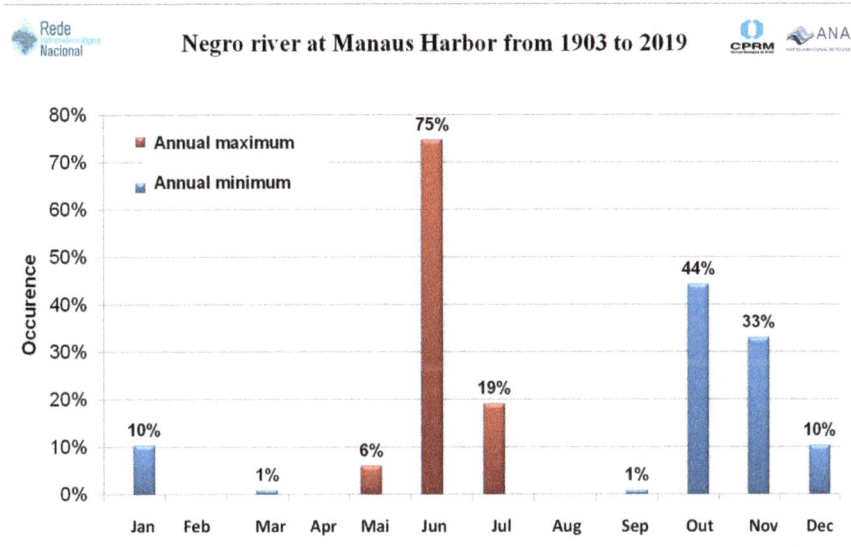

Figure 5:  Monthly distribution of maximum and minimum water levels in Negro River.

Table 2:  Monthly river discharges (m$^3$/s). *(Source: Adapted from CPRM/SGB [12].)*

| Year | Section | River | January | February | March | April | May | June |
|------|---------|-------|---------|----------|-------|-------|-----|------|
| 2008 | Paricatuba | Negro | 24.854 | 19.708 | 24.060 | 22.119 | 29.823 | 61.411 |
| 2009 | Paricatuba | Negro | 38.142 | 41.423 | 30.056 | 35.032 | 34.820 | 37.212 |
| 2010 | Paricatuba | Negro | 8.685 | 10.092 | 9.266 | 20.625 | 45.452 | 58.192 |
| 2011 | Paricatuba | Negro | 23.924 | 20.302 | 33.242 | 31.250 | 55.629 | 64.892 |
| 2012 | Paricatuba | Negro | 24.085 | 27.225 | 38.071 | 44.882 | 62.591 | 55.268 |
| | | | | | | | | |
| 2008 | Manacapuru | Solimões | 85.953 | 103.874 | 108.764 | 118.886 | 127.161 | 125.028 |
| 2009 | Manacapuru | Solimões | 85.301 | 104.337 | 116.431 | 128.542 | 136.241 | 143.218 |
| 2010 | Manacapuru | Solimões | 88.072 | 91.185 | 100.073 | 110.047 | 118.654 | 123.738 |
| 2011 | Manacapuru | Solimões | 63.139 | 81.787 | 92.116 | 105.277 | 124.130 | 125.241 |
| 2012 | Manacapuru | Solimões | 105.641 | 115.078 | 122.247 | 143.086 | 145.726 | 148.177 |

- The year flood impact may be reflected in the following year, as Solimões dam Negro River, the high quotas at the Manaus station in 2009 impacted Negro River flow in 2010, which can be noticed by the flows below average in January, February and March.
- An average flood year, Negro River has increased flow in the months of May and June. Already Solimões has this increase in March and April. Already in an above-average flood year, the flow of Solimões grows in February.

The long-term average (1903 to 2019) of the annual flood peaks is 2,788 cm, with a standard deviation of 115 cm, the historical record was 2,997 cm in 2012. In Fig. 6, we present the graph with the highest water level of evolution in Manaus station. Featured

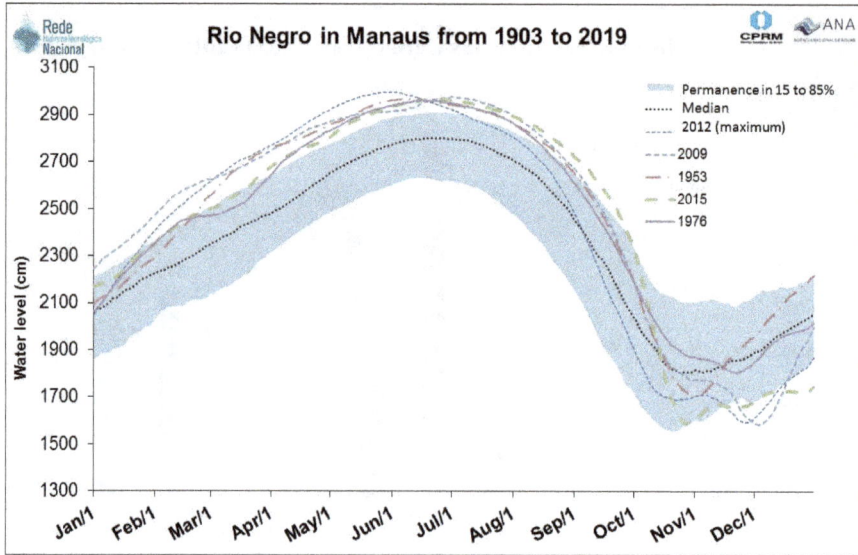

Figure 6: Permanence curve and Negro River largest floods.

maximum height are above the permanence curve which is in the range of 15–85%. Since the maximum water level with 15% of frequency is 2,891 cm.

In Table 3, we present the latest forecasts of the alert and the water level of each year. Each alert forecast has a minimum and maximum range, and an average quota. Thus, in this chart, it is possible to see that in the interval of 15 years, there were 13 hits and two approaches, an efficiency of 87%, which indicates that the forecast model is valid.

Table 3: Alert forecasts and recorded maximum harbour quota in Manaus.

| Year | 1st Forecast March (m) | 2nd Forecast April (m) | 3rd Forecast May (m) | Maximum quota (m) |
|------|------------------------|------------------------|----------------------|-------------------|
| 2019 | 28.49–29.19 | 28.47–29.17 | 29.03–29.33 | 29.42 |
| 2018 | 27.45–28.15 | 27.35–28.05 | 28.20–28.70 | 28.38 |
| 2017 | 29.25–29.95 | 29.15–29.85 | 28.96–29.46 | 29.00 |
| 2016 | 26.60–27.20 | 26.90–27.50 | 26.97–27.57 | 27.19 |
| 2015 | 28.89–29.59 | 28.92–29.62 | 29.39–29.89 | 29.66 |
| 2014 | 28.79–29.49 | 28.84–29.44 | 29.29–29.60 | 29.50 |
| 2013 | 28.75–29.45 | 28.76–29.46 | 29.28–29.71 | 29.33 |
| 2012 | 29.06–29.96 | 29.06–29.96 | 29.97–30.27 | 29.97 |
| 2011 | 27.64–28.34 | 27.78–28.48 | 28.28–28.92 | 28.62 |
| 2010 | 27.08–27.78 | 27.60–28.30 | 27.84–28.53 | 27.96 |
| 2009 | 29.33–30.03 | 29.25–29.95 | 29.15–29.59 | 29.77 |
| 2008 | 28.62–29.32 | 28.08–28.78 | 28.24–28.70 | 28.62 |
| 2007 | 27.08–27.78 | 27.48–28.18 | 27.48–28.18 | 28.18 |
| 2006 | 28.62–29.32 | 28.38–28.88 | 28.73–29.23 | 28.84 |
| 2005 | 28.51–28.91 | 28.35–28.75 | 28.08–28.20 | 28.10 |

Also in Table 3, we can consider:

- In this interval (2005–2019), related to the largest floods, the alert forecast has a similar efficiency (83%), analyzing the first alert, there was a distance, only in 2019.
- Considering in this interval, the largest floods, the alert forecast has a similar efficiency, analyzing the first alert, there was a distance, only in 2019.
- Disclosure of the forecast is important, but what should be considered is the size of the likely flood within the range of acceptable variations. In practical terms, the objective is more qualitative (large, medium, low full) than quantitative accuracy.

## 4  CONCLUSIONS

Negro River levels have been measured since September 1902, with a historical series of 117 years. Predictive studies can be performed using the most diverse statistical and hydrological methodologies. However, we can see that the most important thing is to offer a warning service to the population on the banks and who depend on the river to guarantee their activities. This paper presents selected data that help to understand the hydrology of the region and compose the weekly western Amazon hydrological monitoring bulletin and the three annual alerts in Negro River.

Of the last ten major floods, six are concentrated in this decade, a fact that highlights that these events can become more frequent and thus, knowing the dynamics of rising waters are is something to be conducted by universities and technical sectors that study and operate in the waters from Amazon. The Geological Service of Brazil acts in this monitoring as the sector that collects this primary data and realized the need to disclose the flood forecast and warning project, which presents a possible level range of the Rio Negro for the current year.

As reported, we consider a height of 27 m in Negro River as an alert quota with a return period of 10 years, and when the river reaches a height of 29 m, it is considered an emergency water level, with a return period of 17 years. These levels mean that the waterfront population will feel a kind of impact in their usual activities. Statistical studies, such as linear regression, allows us to obtain a forecast flood level in Manaus from the upstream river stations. In this sense, it is worth noting that the annual flood peaks is 2,788 cm, with a standard deviation of 115 cm, the historical record was 2,997 cm in 2012. Besides, the featured maximum water levels are above the permanence curve which is in the range of 15–85%. Since the maximum height with 15% of frequency is 2,891 cm. However, more important than river level prediction is the estimation of the Manaus flood size within the range of acceptable variations. Even so, our forecasting model has a good result as it is within the flood range and over 80% within the year's maximum height.

This study shows that providing access through roadway infrastructure can provide greater development and better quality of life in rural communities in the Amazon without neglecting the need to control natural resource exploitation and to preserve the environment. Keeping up with the Amazon hydrology is not an easy task, it is dynamic and complex, but understanding the system and the floods evolution, with the slow rise of the waters in approximately 230 days, goes beyond studying the water levels and statistics, is trying to listen and see what nature and the waterfront people try to pass in response to our questions.

## ACKNOWLEDGEMENT
Many thanks to Miguel Arcanjo for the map design and the Geological Survey of Brazil.

## REFERENCES

[1] Chen, J.L., Wilson, C.R. & Tapley, D.B., The 2009 exceptional Amazon flood and interannual terrestrial water storage change observed by GRACE. *Water Resources Research*, **46**, W12526, 2010.

[2] Vale, R., Filizola, N., Souza, R. & Schongart, J., A cheia de 2009 na Amazônia Brasileira. *Revista Brasileira de Geociências*, **4**, pp. 577–586, 2011.

[3] Salati, E. & Vose, P.B., Amazon basin: A system in equilibrium. *Science*, **225**(4658), pp. 129–138, 1984.

[4] Sternberg, H.O., *A água e o homem na várzea do Careiro*, 2nd ed., Museu Paraense Emílio Goeldi: Belém, 330 pp., 1998.

[5] Witkoski, A.C., *Terras, Florestas e Águas de Trabalho. Os camponeses amazônicos e as formas de uso de seus recursos naturais*, Manaus: Editora da Universidade Federal do Amazonas, 486 pp., 2007.

[6] Barichivich, J., Gloor, E., Peylin, P., Brienen, R.J.W., Schöngart, J., Espinoza, J.C. & Pattnayak, K., Recent intensification of Amazon flooding extremes driven by strengthened Walker circulation. *Science Advances*, **4**(9), 7 pp., 2018.

[7] Bittencourt, M.M. & Amadio, S.A., Proposta para identificação rápida dos períodos hidrológicos em áreas de várzea do Rio Solimões-Amazonas nas proximidades de Manaus. *Acta Amazônica*, **37**(2), pp. 303–308, 2007.

[8] Richey, J.E., Nobre, C. & Deser, C., Amazon River discharge and climate variability: 1903 to 1985. *Science*, **246**, pp. 101–103, 1989.

[9] Meade, R.H., Rayol, J.M., Da Conceição, S.C. & Natividade, J.R., Backwater effects in the Amazon River basin of Brazil. *Environmental Geology and Water Sciences*, **18**, pp. 105–114, 1991.

[10] Paiva, R.C.D., Modelagem hidrológica e hidrodinâmica de grandes bacias. Estudo de caso: Bacia do Rio Solimões. Master's dissertation, Universidade Federal do Rio Grande do Sul/Instituto de Pesquisas Hidráulicas, 168 pp.

[11] Marengo, J.A., Borma, L.S., Rodriguez, D.A., Pinho, P., Soares, W.R. & Alves, L.M., Recent extremes of drought and flooding in Amazonia: Vulnerabilities and human adaptation. *American Journal of Climate Change*, **2**, pp. 87–96, 2013.

[12] CPRM/SGB (Companhia de Pesquisa de Recursos Minerais/Serviço Geológico do Brasil), Relatório de Cheia de 2012, Manaus, July 2012. www.cprm.gov.br.

[13] Lima, C., Lall, U., Troy, T.J. & Devineni, N., A climate informed model for non-stationary flood risk prediction: Application to Negro River at Manaus, Amazonia. *Journal of Hydrology*, **522**, pp. 594–602, 2015.

# INTEGRATED MODEL OF MODELS FOR GLOBAL FLOOD ALERTING

BANDANA KAR[1], DOUG BAUSCH[2], JUN WANG[3], PRATIVA SHARMA[4], ZHIQIANG CHEN[4],
GUY SCHUMANN[5], MARLON PIERCE[3], KRISTY TIAMPO[6], RON EGUCHI[7] & MARGARET GLASSCOE[8]
[1]Oak Ridge National Laboratory, USA
[2]Pacific Disaster Center, USA
[3]Indiana University, USA
[4]University of Missouri, USA
[5]Dartmouth Flood Observatory, USA
[6]University of Colorado, USA
[7]ImageCat Inc., California, USA
[8]Jet Propulsion Laboratory, California Institute of Technology, USA

## ABSTRACT

A dramatic increase in frequency of minor to major flooding since 2000 has caused significant economic losses across the world. To mitigate and recover from these losses, actions have been taken to build resilient communities and infrastructures, specifically, by providing situational awareness in near real-time about flood impacts to enhance response and recovery efforts. Several hydrologic and hydraulic flood models are available at various spatial and temporal resolutions to forecast flood events at regional to global scale. Given the global coverage of two operational flood models – GloFAS (Global Flood Awareness System) and GFMS (Global Flood Monitoring System), the purpose of this project is to implement a Model of Models (MoM) approach to integrate the outputs from these two models to classify flood severity at watershed level worldwide, and send alerts based on severity similar to the USGS PAGER (used for severity alerting and impact analysis for earthquakes) to flood impacted communities. The alerts containing flood impacts and severity information will be disseminated through the DisasterAWARE platform, operated by the Pacific Disaster Center (PDC), that provides global multi-hazard alerting and Situational Awareness information to the emergency management community and public. The current version of the MoM approach was implemented for a case study flood event that occurred during January and February of 2020 in South and Central Africa. The findings of the case study event reveal that the approach is effective in identifying potential flood impact areas and the spatio-temporal variation of flood severity, flood depth and extent at watershed level, which will be used to assess infrastructure and societal impacts using earth-observation data and for alerting.
*Keywords: global flood forecasting, flood modelling, model of models, alerting, DisasterAWARE.*

## 1 INTRODUCTION

Flooding is one of the most frequent natural hazards worldwide. Extreme rainfall induced by rapid urbanization and climatic change have contributed to severe flood events, along with significant societal and economic impacts worldwide [1]–[4]. According to the international disaster database (EMDAT) [5], 2019 saw more than 120 floods globally of varying intensity and since 2000, on average, 100 floods annually have occurred worldwide (Fig. 1(a)) that have caused about $10 billion (USD) financial loss per annum (Fig. 1(b)) and significant number of deaths with Asia experiencing the maximum number of fatalities (Fig. 1(c)).

Significant efforts have been made over the past few decades to increase community resilience to flooding and for flood risk management by improving flood risk mapping [6] and forecasting [7], flood impact assessment [8], and floodplain ecology and catchment hydrology [9]. Similar efforts have been made to improve the numerical methods used for flood modelling and use of parallel computing technologies to enhance flood modelling [10]. Irrespective of these modelling advancements, there exists two types of models: (i) empirical and (ii) hydrodynamic [11].

WIT Transactions on The Built Environment, Vol 194, © 2020 WIT Press
www.witpress.com, ISSN 1743-3509 (on-line)
doi:10.2495/FRIAR200071

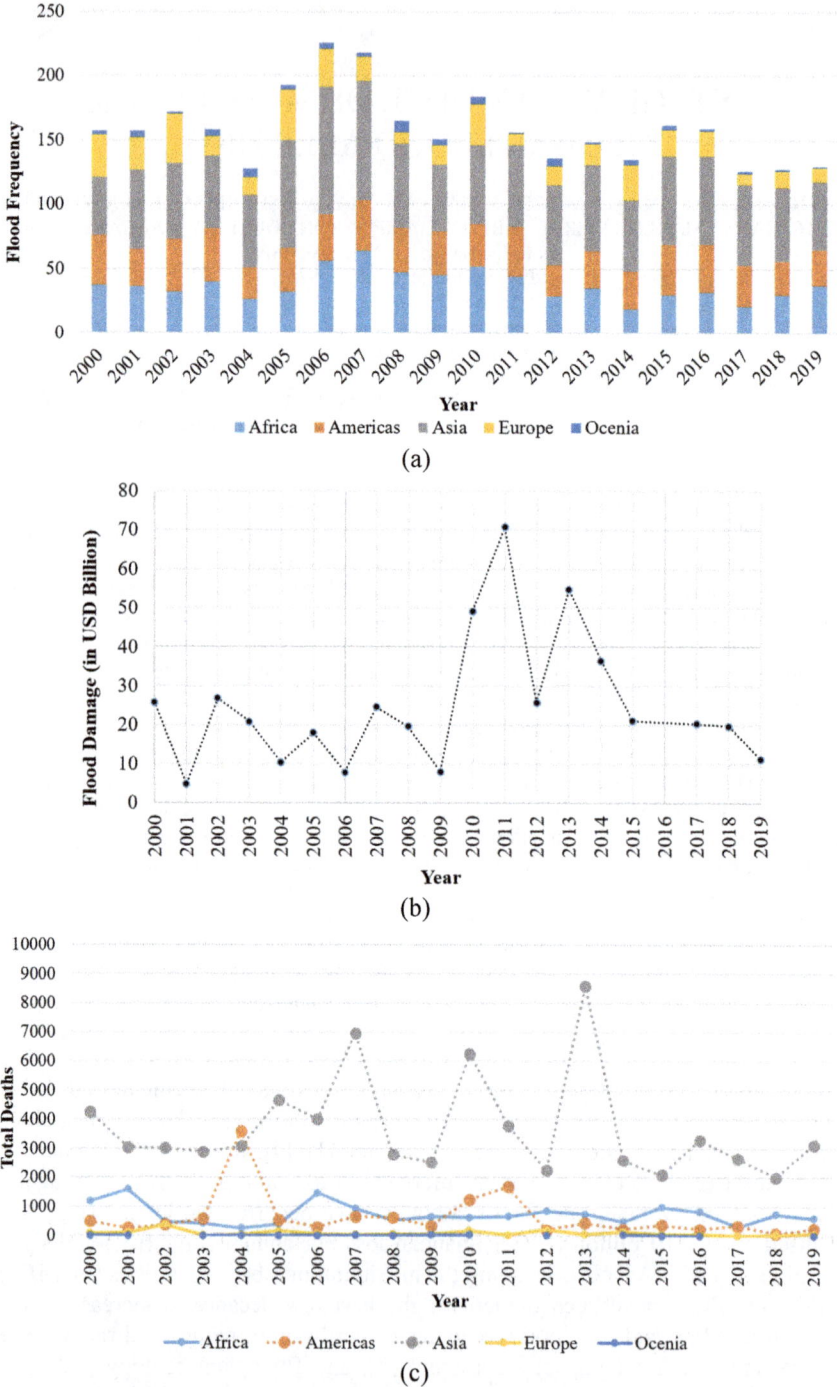

Figure 1:   Global flood statistics 2000–2019. (a) Annual flood frequency; (b) Annual flood induced financial damage; (c) Annual flood induced mortalities. *(Source: Université Catholique de Louvain [5].)*

Empirical models are data driven and developed using historical flood observations, on-ground measurements, surveys, satellite imageries, etc. [11]. Given the nature of inputs used in these models, they are often used for calibration and validation of hydrodynamic models and contain uncertainties. Alternatively, hydrodynamic models are mathematical models that replicate fluid motion of water by accounting for water volume, runoff and soil absorption, topographic conditions, among other things. These models can be one, two and three dimensional depending upon the spatial representation of the flood plain and flood water flow. Because a discussion of these models is beyond the scope of this paper, please refer to Teng et al. [11] for a review of the models.

While empirical models are straightforward, their accuracy is impacted by the spatial and temporal resolutions of the input data, sensor design requirements, environmental factors (cloud cover, weather conditions) as well as statistical approaches used to process the data [11]. By contrast, hydrodynamic models are widely used due to their accurate representation of flood extent and depth. Nonetheless, these models suffer from uncertainties resulting from 1-D, 2D and 3D representation of flood plain and water flow, the input variables and physics used for the models, and the spatio-temporal resolutions of the data sets.

To mitigate and recover from flood induced losses, it is essential to provide situational awareness information to impacted communities in near real-time to aid them with response and recovery efforts, thereby enhancing their resilience. While significant number of flood models are available, these models vary in approach, data and output as well as are implemented at regional to global scale as discussed above. The purpose of this study is to develop and deploy a Model of Models (MoM) approach that integrates two globally operationally flood models – GloFAS (Global Flood Awareness System) and GFMS (Global Flood Monitoring System) – to classify flood severity and send alerts based on severity level to impacted communities similar to USGS PAGER (used for severity alerting and impact analysis for earthquakes) using Pacific Disaster Center's (PDC) DisasterAWARE platform.

The remainder of this paper is organized into five sections. The second section provides a discussion of the DisasterAWARE platform, which is followed by the methodology section that describes the MoM approach used to integrate GloFAS and GFMS and determine flood severity for alerting. The fourth section discusses a case study implementation of the MoM approach using the African flooding that occurred during January–February 2020. The results of the case study are presented and discussed in the fifth section following which the conclusion and future research directions are presented.

## 2 DISASTER-AWARE PLATFORM

DisasterAWARE™ is a system maintained by PDC at University of Hawaii. This system provides multi-hazard warning and situational awareness information for decision support through mobile apps and web-based platforms to millions of users worldwide. Operational version of DisasterAWARE™ is used by multiple national and international agencies including UN agencies for emergency management.

This system continually monitors reliable scientific data sources for events deemed potentially hazardous to people, property, or assets, and posts these incidents as "Active Hazard Alerts." These postings are accessible to decision makers and to the public through early warnings and decision support tools. These Active Hazard Alerts are also available through a Disaster Alert System (DAS) for real-time notifications where users indicate their areas of interest, hazard types and severity for which they would like to receive alerts. While the system provides alerting, severity and decision support for 18 hazard types, potential impact and severity information are provided only for a few hazards (e.g. earthquake, volcano, tropical cyclone) at global scale.

Flooding is a common extreme event, which is also the deadliest as discussed above. However, currently, the DisasterAWARE™ system lacks global flood alerting and does not incorporate a remote sensing component for ground truthing that will allow near real-time validation of flood model prediction outputs. This research focuses on leveraging publicly available and widely used global flood models to determine flood severity.

## 3  METHODOLOGY

To generate an open-access rapid alerting and severity assessment component for global flooding for DisasterAWARE™, a MoM approach is implemented, which integrates existing flood models rather than creating new models. The main features of this model are to: (i) integrate forecasted flood extent and depth information along with flood severity from GloFAS and GFMS, (ii) determine the risk of an area experiencing flood based on severity at global watershed level, (iii) integrate remote sensing observations for validation and calibration of MoM outputs so that actionable knowledge can be generated for decision making. In the following sub-sections, a discussion of GloFAS and GFMS along with the flood severity assessment at watershed level is presented.

### 3.1  Global Flood Awareness System (GloFAS)

The GloFAS system is a global hydrological forecast and monitoring system independent of administrative and political boundaries that is jointly developed by the European Commission and the European Centre for Medium-Range Weather Forecasts (ECMWF) [12]. The system couples state-of-the art weather forecasts with a hydrologic model to provide downstream countries with information on upstream river conditions. GloFAS produces daily flood forecasts and monthly seasonal streamflow outlooks (Fig. 2).

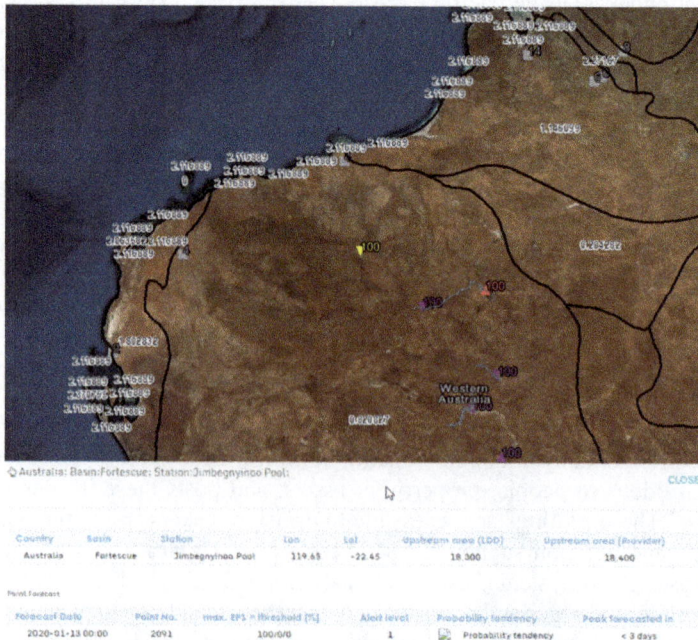

| Country | Basin | Station | Lon | Lat | Upstream area (LDD) | Upstream area (Provider) |
|---|---|---|---|---|---|---|
| Australia | Fortescue | Jimbegnyinoo Pool | 119.65 | -22.45 | 18,300 | 18,400 |

Point Forecast

| Forecast Date | Point No. | max. EPS > threshold [%] | Alert level | Probability tendency | Peak forecasted in |
|---|---|---|---|---|---|
| 2020-01-13 00:00 | 2091 | 100/0/0 | 1 | Probability tendency | < 3 days |

Figure 2:   Output from Global Flood Awareness System. *(Source: European Commission Copernicus Emergency Management Service [12].)*

For the MoM approach, the following hazard severity indicators are obtained from GloFAS: probability of return period events (2, 5 and 20 year), alert level (Medium, High, Severe) and peak forecast (days).

## 3.2 Global Flood Monitoring System (GFMS)

The GFMS uses real-time precipitation information from Tropical Rainfall Measuring Mission (TRMM) Multi-satellite Precipitation Analysis (TMPA) and implements a hydrologic runoff and routing model to produce flood detection/intensity estimates [13]. The system is functional at a quasi-global (50°N–50°S) scale and the hydrologic model is implemented at a 1/8th degree latitude/longitude grid. The flood detection/intensity estimates are based on 13 years of retrospective model runs with TMPA input while flood thresholds are derived at each grid location using surface water storage statistics (95th percentile plus parameters related to basin hydrologic characteristics) (Fig. 3). The model generates streamflow, surface water storage, inundation variables at 1km resolution as well as instantaneous precipitation, and totals from the last day, three days and seven days. For integration in MoM, the following indicators from GFMS are extracted at every 3-hour interval at 0.125 degree grid resolution: size (area and % area in a watershed impacted by a flood), depth above baseline (mean and max) and duration (days).

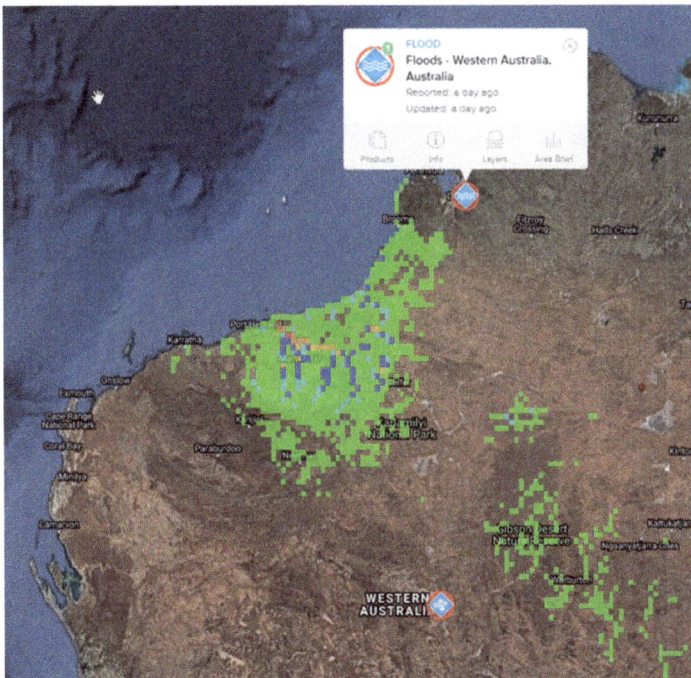

Figure 3:  Global Flood Monitoring System. *(Source: University of Maryland [13].)*

## 3.3 Watershed risk index

For the purpose of global flood severity assessment, the outputs from both GloFAS and GFMS are integrated at a static watershed level. The watershed boundaries are based on hazards rather than political boundaries, which helps generate flood severity and impact information for an entire watershed and population impacted by the flood within the watershed to be used by stakeholders.

Although several watershed datasets are available, the World Resources Institute (WRI) product is used along with the MoM approach as it provides both the geometry information and flood risk attributes (WRI, 2019) for ~3,400 basins across the world. WRI incorporates both the flood risk and drought risk information as well as the baseline and projected future conditions. The following flood risk related attributes are used to determine flood severity at watershed level: Riverine flood risk (rfr_score) and Coastal flood risk (cfr_score).

### 3.4 MoM implementation

The MoM approach implements a cumulative distribution function (CDF) (Fig. 4) to combine the hazard scores determined by integrating GloFAS and GFMS outputs with risk scores at watershed level to compute flood hazard severity at the watershed level. For MoM implementation, first, a weighting approach is used to combine the hazard severity indicators from each model (GloFAS and GFMS) to determine hazard score that ranges between 0–100 (100 being highest score). Second, the watershed risk score (rfr_score or cfr_score) is rescaled from 0–5 to 0–100. Finally, the risk at watershed level and hazard score from the flood models are combined using a CDF to determine the probability of flood at each watershed, which is used to determine dissemination of alert messages based on certain threshold. For instance, based on the probability (derived by using CDF in Fig. 4), a warning will be sent at 75%–100%, a watch advisory will be sent at 50%–75%, an advisory will be disseminated at 25%–50% and just information will be sent at <25%.

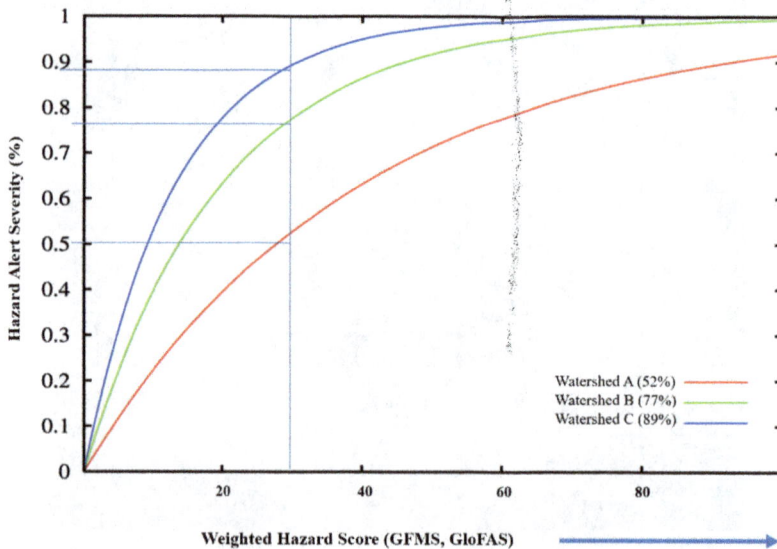

Figure 4:  Watershed risk based on cumulative distribution function.

## 4 CASE STUDY

The MoM approach was implemented for a case study flood event that occurred in early 2020 in Africa. During January and February of 2020, significant number of floods have impacted the African continent [14]. These events have been severe as they have caused a number of fatalities and homelessness in different parts of Africa (Madagascar, Tanzania, Zimbabwe, among others). From emergency management perspective, it is crucial to provide flood alerts for these deadly events to communities in near real-time to aid with evacuation and other response activities. A discussion of the steps implemented to determine flood severity of the African countries using MoM approach on 11th February 2020 is presented below.

## 4.1 Data processing and analysis

For the African flood, the flood depth above threshold (in mm) data were obtained from the GFMS site in binary raster format with 4-byte float type for every 3-hour time step and 0.125° spatial resolution. Using the dimension and geo-reference information for the grids (e.g., row = 800, col = 2458, xllcorner = –127.25, yllcorner = -50 cellsize = 0.125 and, NoDat a = –9999), header files were created for each binary file, which were then used to create Virtual Raster Datasets (readable format within GDAL library). The raster layers and the watershed boundary files were converted to EPSG:3857 WGS84 Web Mercator (Auxiliary Sphere) spatial reference system following which Zonal statistics were performed using watershed boundaries on these raster layers to obtain – flooded area ($km^2$), percentage flooded area (flooded area/watershed area), maximum and minimum flood depth above threshold (mm), and time stamp of flood occurrence.

Flood data for the same event were obtained from GloFAS using its Web Map Service. The output contained points (gage stations) within a specific watershed with the following attributes – country name, watershed basin, station location, upstream area, alert level, peak forecasted days, probability of threshold exceedance based on 2yr, 5yr and 20yr forecasted thresholds. Using Python/GDAL, steps were implemented to automate data download from both flood models.

## 4.2 Flood severity estimation

To estimate flood severity (in percent probability) for the African countries and watersheds, the following steps were implemented:

1. The GloFAS outputs for 11th February 2020 (Alert level, Peak forecast, GloFAS_2yr, GloFAS_5yr, GloFAS_20yr) were rescaled between 1–10 using Table 1.
2. The GFMS outputs for 11th February 2020 (area, % watershed area, mean and maximum depth, and duration (hours the watershed was flooded)) were rescaled between 1–10 as per the weighting scheme for each interval. Because GFMS outputs were generated at 3-hr interval, weighted averages were computed for each output.
3. The weighted indicators were summed together to determine weighted hazard score for impacted watersheds in Africa. The riverine flood risk of each watershed was scaled between 0–100 and incorporated PDC's lack of resilience parameter to identify watersheds with high risk for flood loss due to low resilience.
4. The weighted hazard score (from flood models) was integrated with scaled riverine flood risk at watershed level using CDF to compute severity score (percent probability of experiencing severe flood) (Fig. 5).
5. The computed severity score – a representation of the probability of severe flood impacts – was used to identify watersheds that were at a higher risk of experiencing flooding. The severity score was also classified to identify when alerts and warnings will be disseminated for preparedness and response activities.

Table 1:  Weighting scheme for hazard severity indicators.

| Field | Description | Initial weighting |
|---|---|---|
| GFMS_TotalArea_km | Total impact area in watershed | 1 pt for every 100sqkm, max = 10 (e.g. 890 sq km = 8.9) |
| GFMS_%Area | GFMSArea/WatershedArea | %100/10 (e.g. 66% = 6.6) |
| GFMS_MeanDepth | Mean depth in watershed in mm | 1 pt for every 10 mm, max 10 (e.g. 56 mm = 5.6) |
| GFMS_MaxDepth | Max depth in watershed in mm | 1 pt for every 100 mm, max = 10 (e.g. 890 mm = 8.9) |
| GFMS_Duration | Number of 3-hr intervals a specific area has been flooded (at least 100 square km overlap in each interval) | Continuous days of at least 100 sq km overlap, 1 per day, max 10 (e.g. 66 hours = 2.75) |
| GloFAS_20yr% | EPS threshold % based on 3rd entry (e.g. 86/53/22 = 22%) | %100/10 (e.g. 66% = 6.6) |
| GloFAS_5yr% | EPS threshold % based on 2nd entry (e.g. 86/53/22 = 53%) | %100/10 (e.g. 66% = 6.6) |
| GloFAS_2yr% | EPS threshold % based on 1st entry (e.g. 86/53/22 = 86%) | %100/10 (e.g. 66% = 6.6) |
| GloFAS_AlertLevel | Alert Level 1–3 with 3 greatest value | Level * 3, max 10 (e.g. 3 = 9) |
| GloFAS_PeakForecasted | Number of days until peak forecast arrives at observation point | Weight in days where 1 = 10, 2=9, … 10 or greater = 1 |
| Dynamic hazard score | Updated every 3 hours based on GFMS and 24 hours based on GloFAS | 0–100 sum of hazard inputs |

## 5  RESULTS AND DISCUSSION

Fig. 5 depicts the distribution of the severity score for eight countries in South-Central Africa. Based on the precipitation volume, it is evident from the figure that the Democratic Republic of the Congo (DRC) and Mozambique were the two countries with high risk of experiencing severe flood. According to the PDC's lack of resilience indicators, these two countries also do not have significant resilience initiatives in place, which makes them more susceptible to experiencing significant financial and societal losses from this flood event and future flood events if they occur after February 2020.

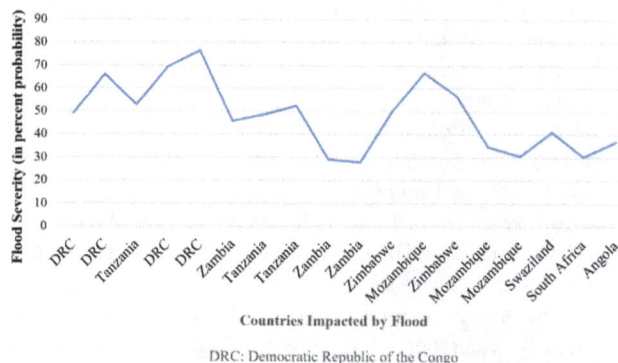

DRC: Democratic Republic of the Congo

Figure 5:   Flood severity score of South and Central African countries on 11th February 2020.

Fig. 6 displays the spatial distribution of the impacted countries and their associated watersheds. Evidently, not all watersheds within a country have similar severity. For instance, the watersheds in DRC have a moderate to very high probability of experiencing severe flood and they also need to receive warnings to prepare for the event. All watersheds in Zambia, Zimbabwe, Tanzania and Mozambique have moderate to high severity, and watersheds in the remaining countries – Angola, South Africa, Swaziland have moderate severity. This severity score is dynamic, which will change with increasing precipitation and flooding, and therefore, can be used to disseminate alerts or watches to these countries in real-time with progress of future floods to help with emergency management efforts.

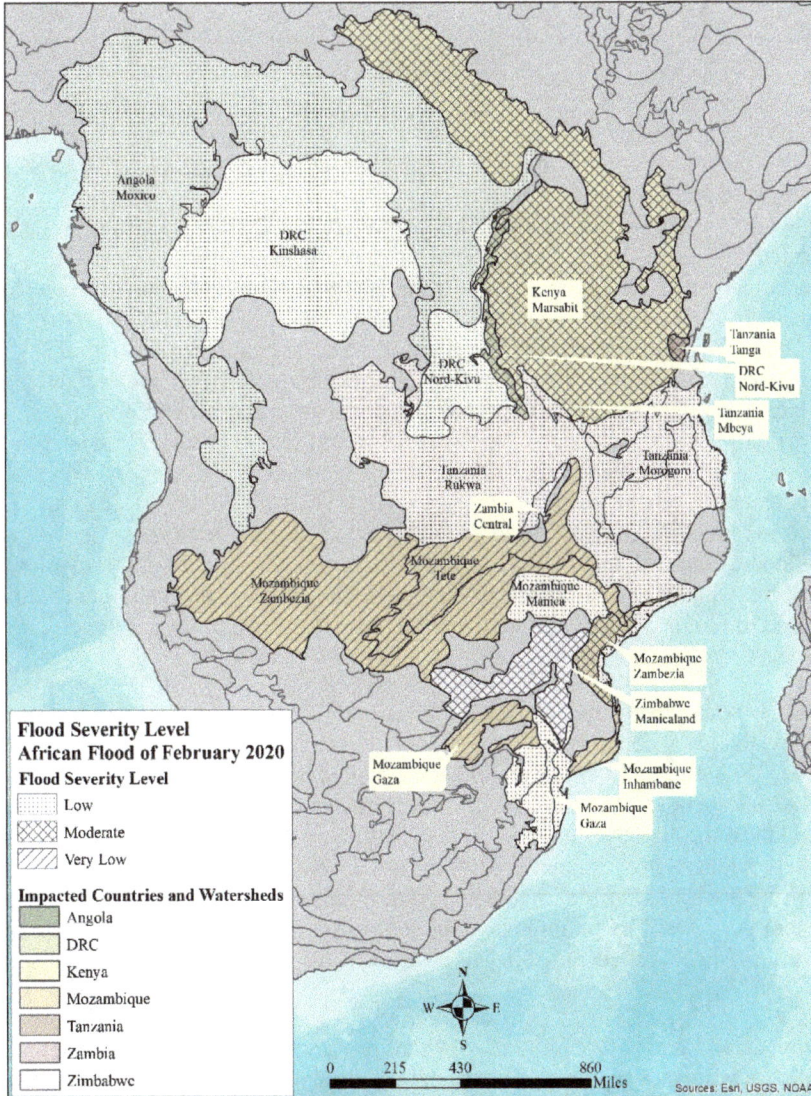

Figure 6:   Flood severity score for South and Central African countries and corresponding watersheds on 11th February 2020.

## 6 CONCLUSION AND FUTURE DIRECTION

Despite the availability of several flood models, the lack of consistency in inputs, outputs, scale of analysis and resolution of data sets as well as the models used to represent terrain and waterflow contributes to generating varying flood outputs to be useful across a broad range of applications. As an alternative, in this project we implemented a MoM approach that integrated the output of real-time flood models at a global scale that has wide application. Evidently, the MoM approach is effective in forecasting flood extent and depth as well as in determining flood severity in impacted areas such that alerts can be disseminated in real-time to impacted communities.

According to the MoM approach, most of the countries in South and Central Africa have moderate to very high severity, which means these countries are at high risk of experiencing severe flood. With continuous precipitation and lack of resilience initiatives in these countries, it can be expected that these countries will experience significant flood induced damages and financial losses, and subsequently, longer time to recover. However, the output of the MoM approach can be used to disseminate alerts and warnings ahead of time to stakeholders and local communities in these countries to prepare effective strategies to mitigate flood induced losses.

In the current implementation of the MoM approach, the weighting scheme was implemented by using the attributes that impact flood severity. In future, a spatial implementation of the weighting scheme using Multi-Criteria Decision Analysis (MCDA) will be conducted that will also account for the time series distribution of the model outputs. The MCDA based weighted approach will ensure that the temporal variation of flooding is accounted for such that the areas experiencing heavy precipitation are weighted highly, thereby increasing their severity and probability of experiencing heavy flood and significant loss. In the current approach, an average was computed for flood severity indicators over a 24-hr period using 3-hr interval GFMS outputs. In future, a spatio-temporal weighting will be implemented to ensure that the flood severity varies for different locations within a watershed across different time span to make flood severity estimation more granular. Finally, flood extent and depth derived from satellite imagery and Digital Elevation Model will be used to validate and calibrate MoM outputs. The validated and calibrated outputs will then be used to estimate flood induced damages, impacted population and losses.

## ACKNOWLEDGEMENTS

The research was carried out at the Jet Propulsion Laboratory, California Institute of Technology, under a contract with the National Aeronautics and Space Administration (80NM0018D0004). This manuscript has been authored by UT-Battelle, LLC under Contract No. DE-AC05-00OR22725 with the U.S. Department of Energy. The U.S. Government retains and the publisher, by accepting the article for publication, acknowledges that the U.S. Government retains a non-exclusive, paid-up, irrevocable, world-wide license to publish or reproduce the published form of this manuscript, or allow others to do so, for United States Government purposes. The findings and opinions presented in this manuscript are those of the authors, and do not reflect the policy or position of any of the aforementioned institutions.

## REFERENCES

[1]  Curebal, I., Efe, R., Ozdemir, H., Soykan, A. & Sönmez, S., GIS-based approach for flood analysis: Case study of Keçidere flash flood event (Turkey). *Geocarto International*, **31**(4), pp. 355–366, 2016.

[2]    Korichi, K., Hazzab, A. & Atallah, M., Flash floods risk analysis in ephemeral streams: A case study on Wadi Mekerra (northwestern Algeria). *Arabian Journal of Geoscience*, **9**(11), p. 589, 2016.

[3]    Yang, L. et al., Structure and evolution of flash flood producing storms in a small urban watershed. *Journal of Geophysical Research: Atmosphere*, **121**(7), pp. 3139–3152, 2016.

[4]    Yin, J., Yu, D., Yin, Z., Liu, M. & He, Q., Evaluating the impact and risk of pluvial flash flood on intra-urban road network: A case study in the city center of Shanghai, China. *Journal of Hydrology*, **537**, pp. 138–145, 2016.

[5]    Université Catholique de Louvain (UCL) – CRED, EMDAT: The Emergency Events Database. www.emdat.be/. Accessed on: 23 Feb. 2020.

[6]    Dutta, D., Herath, S. & Musiake, K., An application of a flood risk analysis system for impact analysis of a flood control plan in a river basin. *Hydrological Processes*, **20**(6), pp. 1365–1384, 2006.

[7]    Arduino, G., Reggiani, P. & Todini, E., Recent advances in flood forecasting and flood risk assessment. *Hydrology and Earth System Science*, **9**(4), pp. 280–284, 2005.

[8]    Bhuiyan, M. & Dutta, D., Analysis of flood vulnerability and assessment of the impacts in coastal zones of Bangladesh due to potential sea-level rise. *Natural Hazards*, **61**(2), pp. 729–743, 2012.

[9]    Karim, F. et al., Assessing the impacts of climate change and dams on floodplain inundation and wetland connectivity in the wet-dry tropics of northern Australia. *Journal of Hydrology*, **522**, pp. 80–94, 2015.

[10]   Vacondio, R., Aureli, F., Ferrari, A., Mignosa, P. & Dal Palù, A., Simulation of the January 2014 flood on the Secchia River using a fast and high-resolution 2D parallel shallow-water numerical scheme. *Natural Hazards*, **80**(1), pp. 103–125, 2016.

[11]   Teng, J., Jakeman, A.J. & Croke, B., Flood inundation modelling: A review of methods, recent advances and uncertainty analysis. *Environmental Modeling and Software*, **90**, pp. 201–216, 2017.

[12]   European Commission Copernicus Emergency Management Service, The Global Flood Awareness System – GloFAS – in a nutshell. www.globalfloods.eu/. Accessed on: 23 Feb. 2020.

[13]   University of Maryland, Global Flood Monitoring System. http://flood.umd.edu/. Accessed on: 23 Feb. 2020.

[14]   FloodList, European system for earth monitoring. http://floodlist.com/africa. Accessed on: 23 Feb. 2020.

# SECTION 3
# FLOOD MODELLING

# TWO-DIMENSIONAL SIMULATION OF NON-NEWTONIAN FLOW

ANDRÉS V. PÉREZ[1*], REYNA HUANCARA[1], FLOR CUTIRE[1], NATALY PEREZ[2] & ANAI PEREZ[3]

[1]School of Civil Engineering, National University of Saint Agustin, Peru
[2]Pratt School of Engineering, Duke University, USA
[3]Graduate School of Architecture, Planning and Preservation, Colombia University, USA

## ABSTRACT

Every year in the south of Peru, rainy seasons produce mud flow along a river in the Andes Mountains, causing flooding in the Pescadores Valley, as well as the Panamericana Highway, paralyzing vehicular traffic along southern Peru's main road. The flooding is of a non-Newtonian flow comprised of torrential rainwater, stones and sediments, this flow will be simulated in the Pescadores valley in order to avoid annual flooding and to propose channeling the river with an embankment dam and a bridge over the Panamericana Highway. For the two-dimensional computational simulation, the hydrological design of the basin was made from the maximum rainfall of 24 hours, until the hydrograph of maximum design flows was obtained, considering a return period that will be based on the expected life of the structure and a percentage of the risk of failure. In addition, a study of soil mechanics was conducted in order to determine the rheological parameters of the flow, as well as a topographic survey in the field with a drone system. With the following information the flow was simulated using the rheological, quadratic and two-dimensional models, as well as numerically solving the continuity equations and quantity of non-Newtonian flow movement (mud flooding). Once the calibration was done, the design of the channelling of the river with an embankment dam and a concrete bridge over the road was proposed.

*Keywords: two-dimensional simulation, mud flooding, hazard map, Peru.*

## 1 INTRODUCTION

Due to global warming, the climate in many regions of Peru has been drastically changing. This study evaluated the intense rainfall that occurs during the summer season throughout the Peruvian mountain range, which cause destructive events known as mud flooding.

Mud flooding are non-Newtonian flows which cannot be studied by river hydraulics making it necessary to use a rheological model. On the other hand, the Pescadores Valley presents an urban topography, therefore, to obtain an accurate model it is indispensable the use of a two-dimensional numerical model. In order to do this, topographic, geotechnical and hydrological studies of the area were first carried out, in order to know the characteristics of the basin that will serve as data in the simulation. Finally, a solution to the interruption of vehicular traffic and the flooding of the valley is proposed.

## 2 PROBLEMATIC

In 2017, due to the hydrological phenomenon of El Niño in Peru, mudflows and floods were produced that destroyed hundreds of kilometers of roads and more than 200 bridges, leaving entire villages isolated and causing the deaths of more than 100 people, causing an economic damage of more than US $3,000 million in damages [1].

The intersection of the Panamericana Highway South at km 756 + 400, which crosses the Pescadores Valley, is frequently affected by sludge avenues, causing frequent blockages, uncommunicated cities, and affecting crops and settlers (Fig. 1).

* ORCID: *https://orcid.org/0000-0002-3404-4349*

WIT Transactions on The Built Environment, Vol 194, © 2020 WIT Press
www.witpress.com, ISSN 1743-3509 (on-line)
doi:10.2495/FRIAR200081

Figure 1:  Panamericana South Highway km 756 + 400, traffic interrupted by mud flooding.

## 3  GOVERNING EQUATIONS

Hydraulics studies Newtonian flows, that is, liquids such as water; however, sludge flows are non-Newtonian flows that are analyzed by rheology.

It is a non-Newtonian flow when the relation between the magnitude of the applied shear stress and the variation of deformation is nonlinear [2].

Mud flow is the mixture of sediments and water in a proportion between 20% and 60% that flows through a channel; in Peru the mud flow is known as "huaicos" [3].

To analyze, formulate, and solve this phenomenon it is necessary to solve the dynamic wave equation in two dimensions. These events present cones of dejection in the valleys, so it is necessary to consider a two-dimensional mathematical model.

In order to properly choose the mathematical model, it is necessary to know the characteristics that the event to simulate presents; therefore, it was decided to use the two-dimensional model that simulates mudflows.

The two-dimensional governing equations [4] correspond to the continuity equations, and these present a numerical solution

$$\frac{\partial h}{\partial t} + \frac{\partial (hu)}{\partial x} + \frac{\partial (hv)}{\partial y} = i \,, \tag{1}$$

where:

$h$ is the depth of mud flow;

$u$ and $v$ are the velocities of the mud flow in x and y directions respectively;

$t$ is time; and

$i$ is the intensity of precipitation.

The equation of quantity of movement is expressed as follows

$$S_{f_x} = S_{0x} - \frac{\partial h}{\partial x} - \frac{u}{g}\frac{\partial u}{\partial x} - \frac{v}{g}\frac{\partial u}{\partial y} - \frac{1}{g}\frac{\partial u}{\partial t} \,, \tag{2}$$

$$S_{f_y} = S_{0y} - \frac{\partial h}{\partial y} - \frac{v}{g}\frac{\partial v}{\partial y} - \frac{u}{g}\frac{\partial v}{\partial x} - \frac{1}{g}\frac{\partial v}{\partial t} \,, \tag{3}$$

where:

$g$ is the acceleration of gravity;

$S_{fx}$ is the slope of friction; and

$S_{0x}$ is the slope of channel.

Non-Newtonian tangential efforts are expressed with the rheological quadratic model, presented by O'Brien and Julien in 1985 [5]

$$\tau = \tau_y + \eta \frac{\partial u}{\partial y} + C_1 \left( \frac{\partial u}{\partial y} \right)^2, \tag{4}$$

where:

$\tau_y$ is the shear stress;

$\eta$ is the dynamic viscosity;

$\dfrac{\partial u}{\partial y}$ is the velocity gradient; and

$C_1$ is the dispersive turbulent parameter.

The procedure for obtaining the dispersive turbulent parameter $C_1$, is the following

$$C_1 = \rho_m l^2 + f(\rho_m, C_v) d_s^2. \tag{5}$$

In these equations, $\eta$ is the dynamic viscosity; $\tau_c$ is the cohesive yield stress, $\tau_{mc}$ is the Mohr Coulomb stress which depends on the intergranular pressure (ps), and the angle of repose ($\varphi$) of the material; $C_1$ denotes the initial shear stress coefficient, which depends on the mass density of the mixture $\rho_m$, the Prandtl $l$ mixture length, the size of the sediment ds and a function of the volumetric concentration of sediments $C_v$. Bagnold [6] defined the function

$$f(\rho_m, C_v) = a_i \rho_m \left[ \left( \frac{C*}{C_v} \right)^{1/3} - 1 \right], \tag{6}$$

where:

$a_i$ is the empirical coefficient ($\sim 0.01$); and

$C*$ is the maximum static volume concentration of sediment particles.

## 3.1 Numerical solution of the FLO-2D model

The shear stress in hyperconcentrated sediment flows can be determined by the sum of the five shear stress components. The total shear stress $\tau$ depends on the cohesive yield stress $\tau_c$, the Mohr–Coulomb stress $\tau_{mc}$, the viscous stress $\tau_v$ ($\eta \, dv/dy$), the turbulent stress $\tau_t$ and the dispersive stress $\tau_d$

$$\tau = \tau_c + \tau_{mc} + \tau_v + \tau_t + \tau_d, \tag{7}$$

where:

$$\tau_y = \tau_c + \tau_{mc}. \tag{8}$$

To define all the shear stress terms used by the FLO-2D model, the shear stress ratio is integrated in depth and is rewritten as a dimensionless slope as follows:

$$S_f = S_y + S_v + S_{td}, \tag{9}$$

where:
$S_f$: slope of friction;
$S_y$: slope of creep;
$S_v$: viscous slope; and
$S_{td}$: turbulent-dispersed slope.

The viscous slope and the turbulent-dispersed slope are written in terms of depth-averaged velocity $V$. The viscous slope can be written as:

$$S_v = \frac{K\eta}{8\gamma_m} \frac{V}{h^2},\tag{10}$$

where:
$\gamma_m$ : specific weight of the sediment mixture; and
$K$: laminar flow resistance parameter.

The flow resistance $n_{td}$ of the components of the turbulent and dispersive shear stress are combined as an equivalent Manning value n for the flow:

$$S_{td} = \frac{n_{td}^2 V^2}{h^{4/3}}.\tag{11}$$

At very high concentrations, the dispersive stress reached by the contact of the sediment particles increases the flow resistance $n_{td}$ by the transfer of impulse flow to the boundaries. To estimate this increase in flow resistance, the conventional turbulent resistance flow $n_t$ is increased by an exponential function of the sediment concentration $C_v$ (O'Brien and Julien [5])

$$n_{td} = n_t b e^{mC_v},\tag{12}$$

where $b$ is a coefficient (0.0538) and $m$ an exponent (6.0896).

The components of the friction slopes are then combined in the following way:

$$S_f = \frac{\tau_y}{\gamma_m h} + \frac{K\eta V}{8\gamma_m h^2} + \frac{n_{td}^2 V^2}{h^{4/3}}.\tag{13}$$

FLO 2D provides a solution with the finite difference method, using an explicit scheme. For the numerical stability of the calculations, uses the Courant–Friedrich–Lewy condition.

The FLO-2D model will solve the equation of momentum to calculate velocity.

## 3.2 Grid characteristics

This model uses a quadrangular structured grid. The topography of the surface is discretized in a grid formed by cells or square elements of uniform size for the entire study area and each element is assigned a position in the grid, an elevation or ground level and a roughness coefficient n of Manning.

The grid cell size should reflect the level of detail of the topographic data, for which a balance should be sought between the grid cell size and the time it takes for the computer to simulate with that cell size.

The cell size used in the present study was 3.00 m, with a resulting number of cells of 353,065.

## 4  BASIC STUDIES FOR THE SIMULATION

### 4.1  Topography

After conducting a previous field survey, the topography survey of the dejection cone of the valley called Los Pescadores was carried out using an eBee RTK drone from a geodetic point provided by the Geophysical Institute of Peru [7] in the Datum system Reference: World Geodetic System (WGS-1984) with Universal Transversal Mercator Cartographic Projection (UTM). Contours were obtained every meter.

Additionally, the differential Global Position System was used to locate and materialize the control points.

### 4.2  Hydrology

The hydrograph was obtained from maximum 24-hour rainfall data due to that these records were the only ones available. The hydrological model HEC-HMS and the Soil Conservation Service (SCS) methodology were used to determine the hydrograph. The 24-hour maximum rainfall records were obtained from the National System of Meteorology and Hydrology of Peru [8]. The net precipitation rate was obtained through the curve number. Then the histogram was transformed into a hydrograph using the unit hydrograph method. The hydrograph is shown in Fig. 2.

The parameters of the HEC-HMS model that were used are those shown in Table 1.

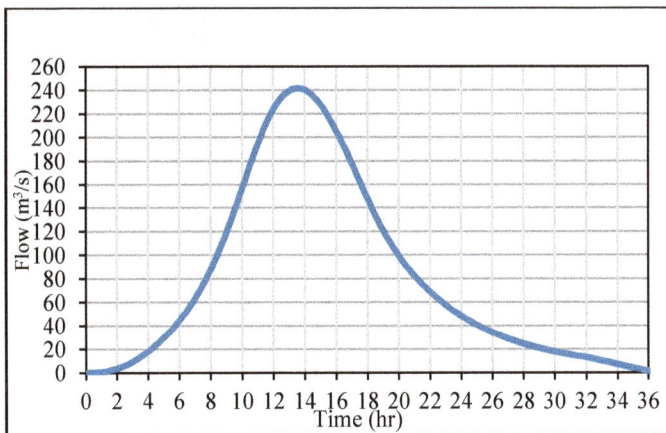

Figure 2:  Hydrogram for 200-year return time.

Table 1:  Parameters of the model HEC-HMS.

| Area (km$^2$) | 1 954.90 |
|---|---|
| Loss method | SCS curve number |
| Transform method | SCS unit hydrograph |
| Baseflow method | None |
| Routing method | Muskingum |
| Curve number | 81.70 |
| Lag time (min) | 1,483.00 |

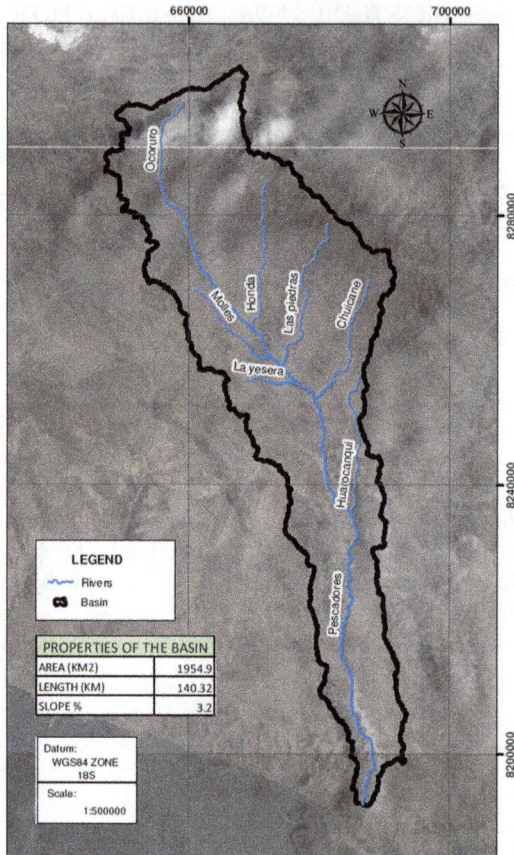

Figure 3:  Caravelí – Pescadores hydrological watershed.

Finally, the hydrograph transit was performed using the Muskingum–Cunge method. Fig. 3 shows the study watershed. It was determined that the maximum water flow for a return period of 200 years is 240 m$^3$/s.

### 4.3  Soil mechanics

Soil samples were taken in different locations of the Pescadores Valley and the riverbed, in order to determine the rheological parameters for the mud flow. The sample that best represented the flow behavior was taken at the beginning of the ejection cone. The soil classification tests determined that the sample corresponds to a sand and silt of low plasticity.

### 4.4  Rheological parameters

The values of rheological parameters such as viscosity and initial stress were calculated. These values were determined by comparing samples collected in situ with "type" samples proposed by O'Brien and Julien in 1988 [9].

After analyzing the samples it was concluded that the sample collected in situ corresponds to the type "Glenwood 3" (Fig. 4).

Figure 4:   Cleaning of material with front loader and tractor, occurred on 12 February 2012 (2:50 pm) [10].

The parameters mentioned above, according to their type of classification, are expressed according to the concentration of sediments. Those expressions are described below:

$$\eta = 0.00632e19.9Cv, \tag{14}$$

$$\tau_y = 0.000707e29.8Cv. \tag{15}$$

In the present study, the volumetric concentration ($Cv$) is 0.35, after the calibration of the model.

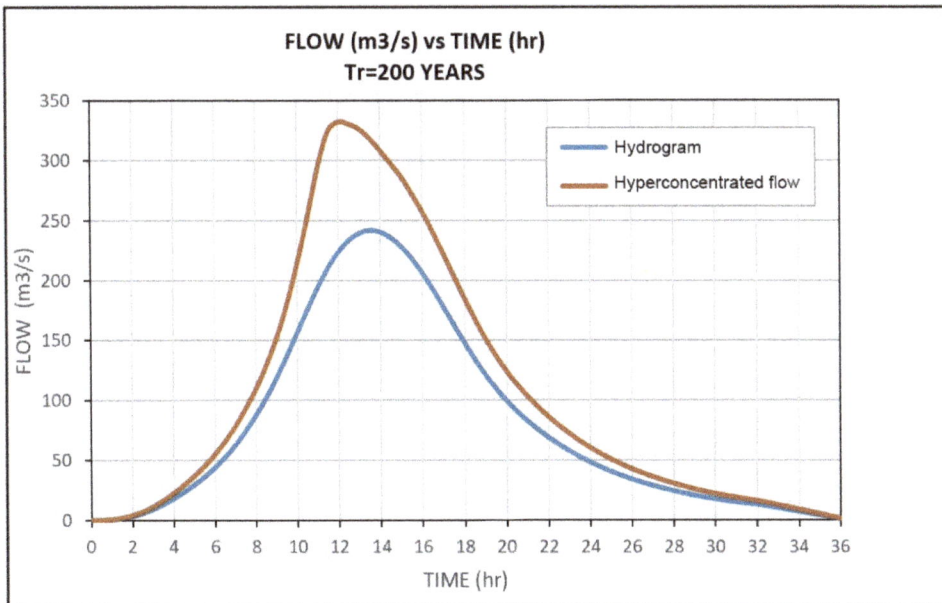

Figure 5:  Hydrogram of hyper concentrated flow for a return period of 200 years.

## 5  MUD FLOODING SIMULATION

The information obtained from the field, such as the topography, hydrology, roughness of the riverbed and soil study, were the input data to FLO 2D which simulated mud flooding. The total hydrograph of water plus mud was also calculated since it is also used for the simulation. It can be seen in Fig. 5.

### 5.1  Calibration of the model

The calibration of the simulation was performed using the maximum precipitation event recorded on 12 February 2012, of which records and audiovisuals are available [11], [12].

For the calibration of the model the estimated in situ flow brace was compared with the brace reported by the model, this approximation was obtained by varying the sediment concentration parameter. The result of this process is shown in Fig. 6.

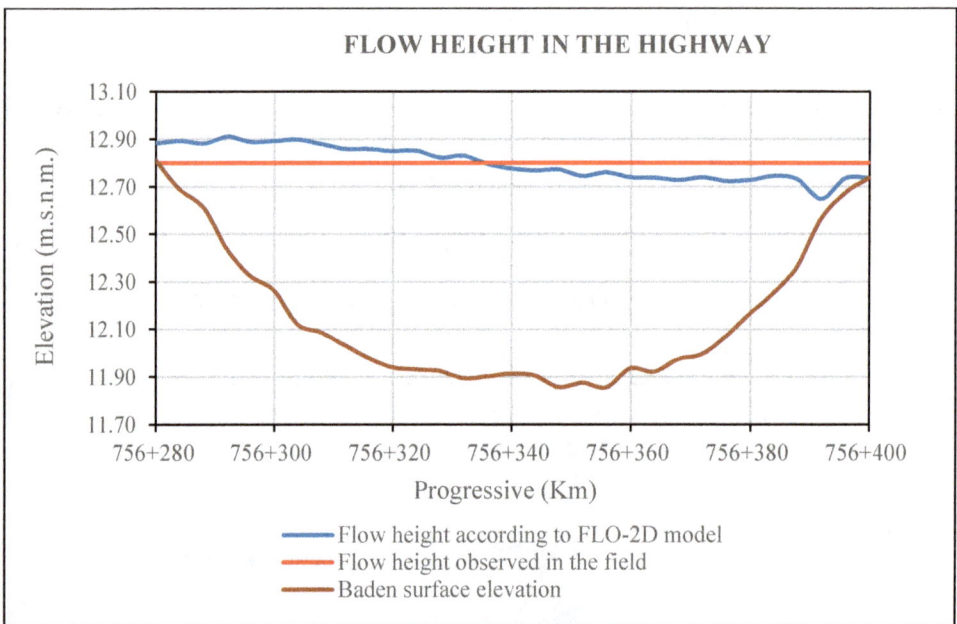

Figure 6:  Comparison between flow observed in situ and simulated after calibration.

In Fig. 6, the corresponding profile is shown to the intersection of the highway with the river, the flow level according to the FLO-2D model is calibrated with respect to the flow level observed in the field.

The event that occurred on 12 February 2012, reached a span of 1.00 m in the center of the ditch, with this value the FLO-2D model was calibrated.

After many iterations it was determined that the sediment concentration is 0.35, according to that value it is classified as a "mud avenue".

Finally, it was determined that the peak flow of the mud flow hydrograph is 332 m$^3$/s as can be seen in Fig. 5.

# 6 RESULTS

## 6.1 Hazard map

The methodology that was followed to graph the hazard map of the Pescadores Valley for urban planning was the one proposed by Garcia et al. [13], this method makes a distinction between water flow and mud flow.

Likewise, this methodology establishes the flow intensity, in terms of the product of maximum flow height h (m) and maximum speed v (m³/s) which are shown in Table 2.

Table 2:  Intensity level of hazard in mud flows.

| Intensity level | Flow height h (m) | Product of maximum flow height h and maximum speed v (m²/s) |
|---|---|---|
| High | h > 1.00 m | vh > 1.00 m²/s |
| Medium | 0.20 m < h < 1.00 m | 0.20 m²/s < vh < 1.00 m²/s |
| Low | 0.20 m < h < 1.00 m | vh < 0.20 m²/s |

Figure 7:  Hazard map in the Pescadores urban area for a 200-year return period.

Fig. 7 shows the hazard map in the case a mud avenue occurs in the Pescadores Valley. This map will allow the local administration a better planning in urban development and zoning.

## 6.2 Mud flooding

After the calibration was performed, a flood simulation of the valley and Panamericana highway was conducted. Those simulations were made for different return periods. This paper presents the one for 200 years [14].

Following the simulation, different solution alternatives were simulated and it was found that the best one is to build a bridge of 105 m in length at the junction of the Pescadores ravine with the Panamericana Sur Highway, as well as the channeling of the valley by means of earth dams and embedded to protect the crops and the Pescadores Valley; these being the most economical, viable and efficient measures.

Fig. 8 displays the comparison between the simulation of a mud avenue for a 200-year return period and the simulation with the proposed solutions (channeling and bridge). In order for an eventual mud flow does not interrupt the transit of the Panamericana Sur Highway.

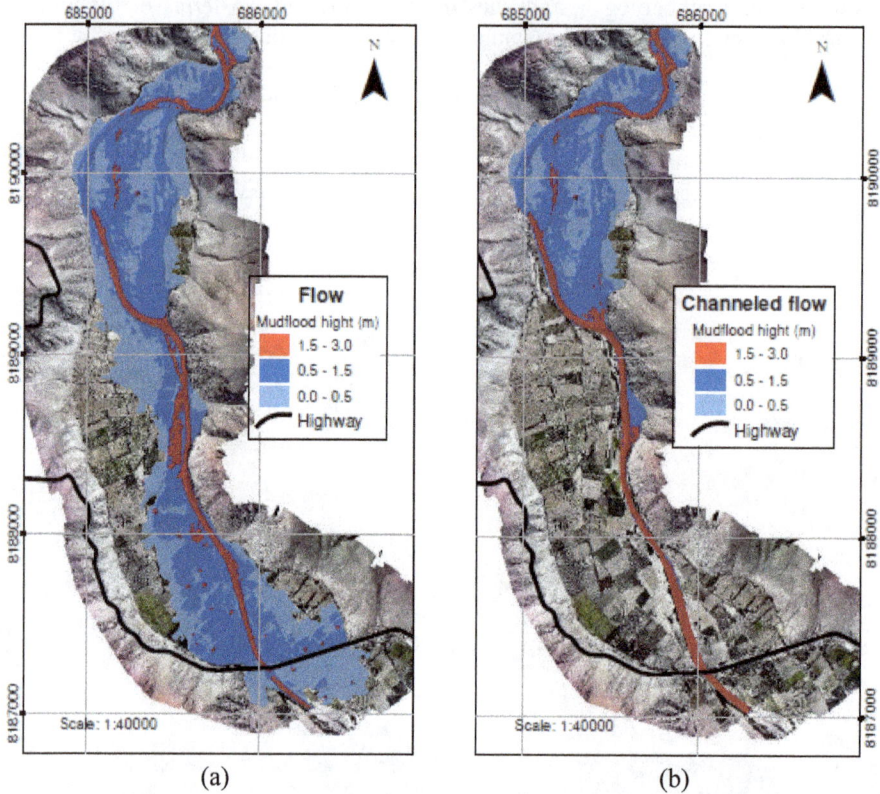

Figure 8:   Depth resulting from the simulation of mud flooding, for a return period of 200 years. (a) No proposed solution; and (b) With proposed solution (bridge and embankment).

The left side of the Fig. 8 presents the simulation without considering the suggested solutions. It can be seen that the mud floods the road and interrupts vehicular traffic and affects the surrounding crops. The right side shows the simulation with the suggested solutions, which as we can see, does not interrupt vehicular traffic.

6.3  Bridge and embankment

The proposed solution considers the construction of a beam-type bridge with a total length of 105 m and a width of 12 m. The channeling must be carried out by means of earthen and embedded dikes, which has a crown width and height of 4.00 m with an average height of 3.0 m. Fig. 9 displays the proposed solutions.

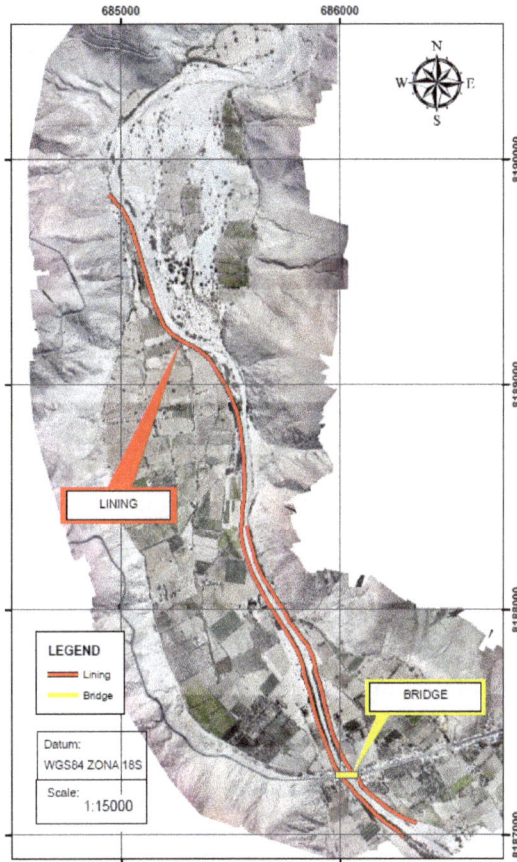

Figure 9:   Proposal for the land, embankment solution and the construction of the Pescadores bridge.

## 7 CONCLUSIONS AND RECOMMENDATIONS

From in situ data it was found that the sediment–water concentration was 0.35, which is classified as a "mud avenue"; from those conditions the simulation was conducted. A suitable solution given the conditions described above would be to channel the flow with earth and rock dikes.

When implementing the proposed solution to the simulation it was verified that the flow would no longer cause floods in the Pescadores valley and would not interrupt traffic in the Panamericana Highway.

Mud flooding simulations need a plethora of in situ information, such as topography, hydrology, soil mechanics, rheological parameters, the necessary knowledge of mathematical models that suit better to the phenomenon, and the numerical solution in order to be able to simulate and obtain optimal results.

It is recommended to install flow capacity stations in strategic points of the river since there is a lack of those. These stations would provide better tools when trying to calibrate simulation models.

Additionally, it is advisable to use two-dimensional models of sludge flows since in its path there is a sludge cone. It would be better to use three-dimensional models; however, their formulations and solutions are almost impossible given their complexities. Future studies will begin to dig deeper into their solutions.

## REFERENCES

[1]  Macro consult, sistema de información. https://sim.macroconsult.pe/danos-de-el-nino-us3-124-millones-hasta-ahora-macroconsult/. Accessed on: 2 Mar. 2019.
[2]  Streeter, V.L., Wylie, F.B. & Bedford, K.W., *Fluid Mechanics*, 9th ed., McGraw-Hill: Colombia, 2000.
[3]  Castillo, N. & Leonardo, F., *Aplicación de un modelo numérico de flujo de escombros y lodo en una quebrada en el Peru.*, Universidad Nacional de Ingeniería, Lima, Peru, 2005.
[4]  O'Brien, J., Julien, P. & Fullerton, W.T., Two-dimensional water flood and mudflow simulation. *Journal of Hydrologic Engineering, ASCE*, **119**(2), 1993. https://doi.org/10.1061/(ASCE)0733-9429(1993)119:2(244).
[5]  O'Brien, J.S. & Julien, P.Y., Physical processes of hyper concentrated sediment flows. *Conference on the Delineation of Landslides, Floods and Debris Flow Hazards*, Utah, USA, pp. 260–279, 1985.
[6]  Bagnold, R.A., An empirical correlation of bedload transport rates in flume and materials rivers. *Proceedings of the Royal Society of London, Series A Mathematical and Physical Sciences*, **372**(1751), pp. 453–473, 1980.
[7]  IGN, Ficha Tecnica PCR-3, Peru, 2017.
[8]  SENAMHI, Descarga de datos hidrometeorológicos, 9 Apr. 2017. www.senamhi.gob.pe/?p=descarga-datos-hidrometeorologicos.
[9]  O'Brien, J.S. & Julien, P.Y., Laboratory analysis of mudflow properties. *Journal of Hydraulic Engineering, ASCE*, **114**(8), 1988. https://doi.org/10.1061/(ASCE)0733-9429(1988)114:8(877).
[10]  PROVIAS, Ministerio de transportes y comunicaciones, Unidad zonal XIV, Arequipa, Report 13, 2012.
[11]  Castillo, A., www.youtube.com/watch?v=Wd2xFrrTdSA. Accessed on: 12 Jan. 2018.
[12]  Zegarra, H., Pasando el huayco por Pescadores panamericana sur. www.youtube.com/watch?v=2BEhPxUlpO8. Accessed on: 15 Mar. 2019.
[13]  Garcia, R., Lopez, J., Noya, M., Bello, M., Gonzalez, N., Paredes, G. & Vivas, M., Mapas de riesgo para eventos de flujo de barro y detritos en el estado de Vargas y Caracas. Informe proyecto Ávila: Caracas, Venezuela, 2002.
[14]  Perez, A.V., Perez, N. & Perez, A., Flooding simulation and channeling in the valley of the Andes mountain in the south Peru. *WIT Transactions on Ecology and the Environmental*, vol. 239, WIT Press: Southampton and Boston, pp. 287–293, 2019.

# ESTIMATION OF FREQUENT PEAK FLOOD DISCHARGE FOR THE UPPER RAJANG RIVER BASIN IN SARAWAK, MALAYSIA

JERRY BETIE CHIN[1,2], KHAMARUZAMAN WAN YUSOF[1] & MUBASHER HUSSAIN[2]
[1]Departmental of Civil and Environmental Engineering, Universiti Teknology Petronas, Malaysia
[2]Hydro Department, Sarawak Energy Berhad, Malaysia

ABSTRACT

Rajang River Basin (RRB) is the largest river basin in Malaysia, in the central region of Sarawak Malaysia. There are two large dams (Murum Dam and Bakun Dam) in upper RRB and these are built as a cascade. Bakun Dam is located downstream of the Murum Dam and to assess the flood risk to the Bakun Dam, the estimation of frequent peak flood discharge (PFD) is important. Rainfall-runoff routing modelling was undertaken with RunOff Routing on Burroughs (RORB) tool to estimate PFD for 1 in 2 annual exceedance probability (AEP) up to 1 in 100 AEP. RORB tool was used to derive flood hydrograph, and a hydrological model was established for the Bakun catchment. Based on the analysis, for the 1-day storm, the 2-year and 100-year return period design rainfall are 115 mm and 206 mm, respectively. For the 3-day storm, the 2-year and 100-year return period design rainfall are 188 mm and 344 mm, respectively. The peak flood discharge for 1-day storms is higher than the 3-day storms. For the 1-day storm, the 2-year and 100-year return period, peak flood discharges are 3,867 m$^3$/s and 7,043 m$^3$/s, respectively. For the 3-day storm, the 2-year and 100-year return period, peak flood discharges are 3,632 m$^3$/s and 6,722 m$^3$/s, respectively.

*Keywords: flood risk assessment, Rajang River Basin, peak flood discharge, RORB, flood frequency analysis.*

## 1 INTRODUCTION

Floods are the most frequently occurring hazard events among all-natural disasters in South East Asia, including the upper Rajang River Basin (RRB) in Sarawak [1]. Flood cause a loss of lives and destruction of properties [2]. RRB is contributing about forty per cent of Sarawak state area, and flood risk assessment for RRB is critical to be assessed for the flood mitigation. RRB is located at the central part of Sarawak; the climate is wet with high rainfall intensity, and because of its topography, it exposes the structures along the Rajang River to the flood risks.

Peak flood discharge (PFD) or design floods are commonly used to assess the potential flood risks. PFD is required for the planning and design of water resources and civil structure projects, flood plain risk management and regulatory compliances. The two commonly used methods for the PFD estimation are streamflow based and rainfall-based [3]. When adequate streamflow data is accessible for a particular site of interest, PFD can be estimated by performing frequency analysis of the recorded inflow data. However, this is not the case for this study, as there is no long-established streamflow station and data. Thus, a rainfall-based flood model will be used, utilising RORB hydrological tool which incorporates the rainfall-runoff routing model.

The RORB tool is commonly used for flood estimation in the Asia Pacific. RORB is an interactive non-linear distributed runoff and streamflow routing program. It calculates catchment losses and streamflow hydrographs from rainfall events input by the user. The hydrology models from RORB were calibrated with the observed stream flows at Bakun Dam.

WIT Transactions on The Built Environment, Vol 194, © 2020 WIT Press
www.witpress.com, ISSN 1743-3509 (on-line)
doi:10.2495/FRIAR200091

Design rainfall storm intensity for the RORB modelling was derived from the rainfall data from six (6) rainfall stations using the rainfall frequency methods. The standard statistical method, Log Pearson Type III, Log-Normal and Gumbel distribution were used to analyse the rainfall frequency.

The key objectives of the study are: (1) to estimate rainfall frequency and intensity; and (2) to estimate the frequent PFD at Bakun Dam.

## 2 STUDY AREA AND DATA DESCRIPTION

### 2.1 Study area

The RRB is the largest river basin in Sarawak, draining an area of approximately 50,000 km$^2$ [1]. The uppermost part of the Rajang River Basin (RRB) which feed into Bakun Dam comprises of four river tributaries, namely Balui River, Bahau River, Linau River and Murum River. Murum River is regulating by the Murum Dam, and the other three river tributaries are freely flowing into Bakun Dam. Fig. 1 shows the overall RRB in Sarawak and Fig. 2 shows the Bakun catchment.

### 2.2 Rainfall data

Historical daily precipitation for six rainfall stations in Bakun catchment was obtained from the Department of Irrigation and Drainage (DID) [4], Sarawak for the period of 1976–2019, as shown in Fig. 2.

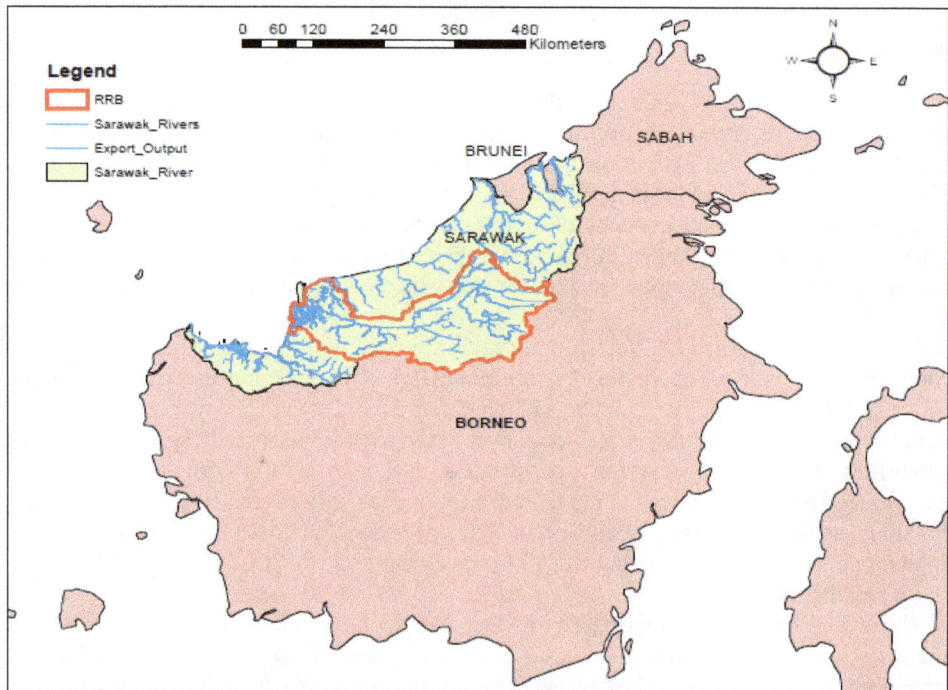

Figure 1: Rajang River Basin in Borneo.

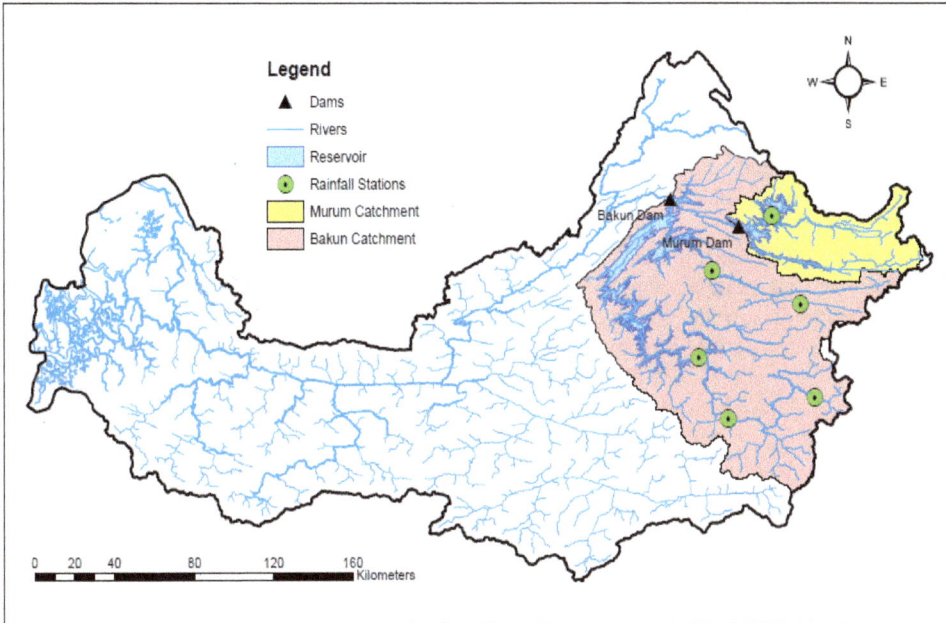

Figure 2:  Murum and Bakun catchment with six (6) rainfall stations.

## 2.3  Water level and inflow data

Sarawak Energy Berhad (SEB) has hydrometric stations at Murum Dam and Bakun Dam which records water level and inflow in Murum and Bakun catchments. Water level and inflow at both dams are being archived in the hydrological database through the telemetry system. Inflow data has been scrutinised to establish its suitability for calibration and validation of the continuous hydrological model in RORB.

## 2.4  Topographic data

The topographic data for the catchment and river reaches were obtained from Hydro Department of Sarawak Energy for the two sources as below:

- Light Detection and Ranging (LiDAR) survey data;
- Interferometric Synthetic Aperture Radar (IFSAR).

### 3  METHODOLOGY

## 3.1  RORB model development

RORB is an interactive, non-linear distributed runoff and inflow routing model. This software was chosen because of its widely recognised capabilities in flood routing. It is also easier to use and not required many inputs that lead to fewer assumptions made for the development of the hydrological model. It has also been widely used in many catchments studies in the Asia Pacific as well as in Malaysia for flood risk assessment. RORB is free software and can be downloaded from the internet.

### 3.1.1 Hydrological modelling

RORB software will be used to perform rainfall-runoff routing modelling to estimate frequent peak flood discharges for 2-year, 5-year, 10-year, 50-year and 100-year return periods.

A RORB model established for the overall Murum and Bakun catchment [5] to estimate frequent peak flood discharge at Bakun Dam. Design rainfalls for the RORB modelling were obtained from the rainfall frequency analysis as described in Section 3.2.1.

### 3.1.2 Catchment delineation and reaches and nodes

RORB model setup begins with the catchment modelling. The catchment was divided into 114 sub-catchments, river links and nodes were connected utilising a combination of LiDAR and IFSAR survey data of the study area. The formed series of link and nodes represent the reaches of flow and the nodes of each sub-catchment, as shown in Fig. 3. Parameters like river length and sub-catchment area were defined and determined.

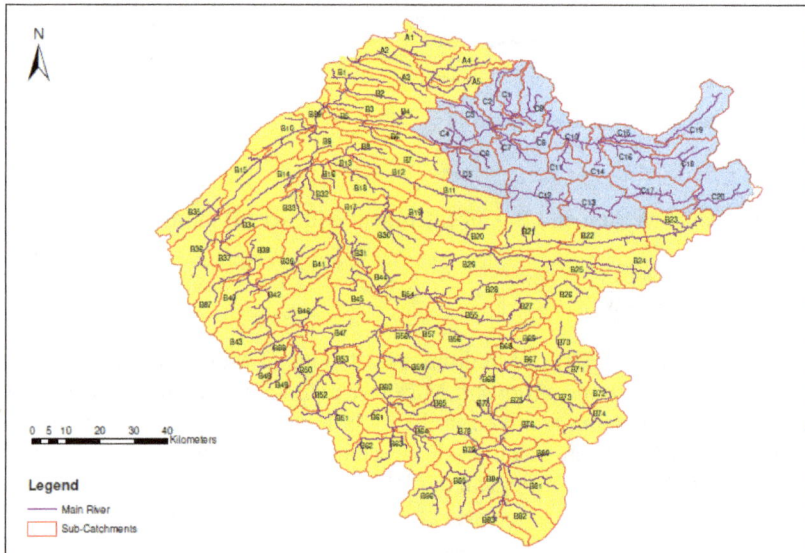

Figure 3:  RORB model layout.

The storage discharge relationship in RORB model is:

$$S = 3600 * K_c * K_{ri} * Q^m, \qquad (1)$$

where $S$ = storage in reach (m$^3$), $Q$ = discharge (m$^3$/s), $K_c$ and $m$ are main catchment parameters that can be obtained through trial and error fitting, known as a fit run in RORB model setup, while $K_{ri}$ = relative routing lag parameter for the specific reach and storage [5].

### 3.1.3 $K_c$ and m values in RORB

In the RORB model, catchment lag and non-linearity are controlled by the factors $K_c$ and $m$, respectively. The selection of $K_c$ and $m$ values were performed through the fit run. Wide range of peak discharges of the recorded inflow was selected to estimate the $K_c$ and $m$ values. Trial and error methods were performed with adjusting range of $K_c$ and $m$ values to obtain the best fit of the recorded inflow hydrograph. The selected pair of $K_c$ and m values were tested against other peak discharge graphs and provided acceptable fit with accuracies within +/–15% [6], which is considered as good accuracy. Among all the fit runs, the m value of 0.75, $K_c$ value of 178 and 0 initial loss (IL) were adopted as the best fit for all the storm events in the study area as shown in Fig. 4.

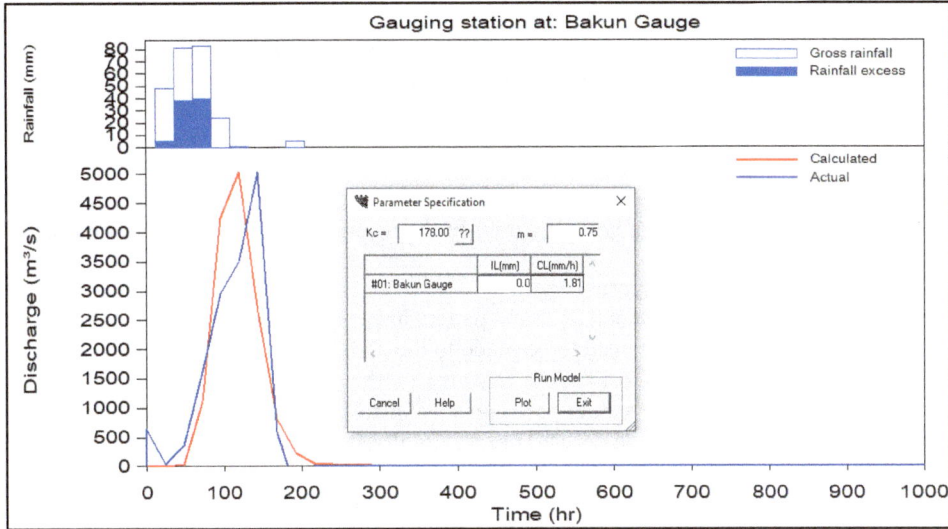

Figure 4:  RORB fitted hydrograph using $K_c$ = 178.

## 3.2  Design rainfall estimate

### 3.2.1  Rainfall frequency analysis

For this study, three frequency distributions Log Pearson Type III, Log-Normal, and Gumbel were adopted and compared for their ability to fit maximum rainfall values in Murum-Bakun catchment using eqns (2)–(5). Analyses were performed for storm periods of 1-day and 3-day. Daily rainfall data from six stations for the 44-year period (1976–2019) provided the basic data for the study.

In Log Pearson Type III, the coefficient of skewness is calculated using eqn (2) as given below:

$$C_s = \left(\frac{1}{\sigma^3}\right)\left\{\frac{N}{(N-1)(N-2)}\right\}\sum (X_i - X_{avg})^3.$$

$$(2)$$

The frequency factor k is obtained from the available theoretical table for the Pearson type III distribution.

A Log-Normal distribution is a probability distribution of a random variable whose logarithm is normally distributed. The maximum rainfall for a specific return period is calculated using eqns (3) and (4) [7]

$$X_T + X_{av} = k\sigma, \qquad\qquad (3)$$

where $X_{av}$ is the mean values, k is the frequency factor and,

$$\sigma = \left[ \frac{\sum (X_i - X_{avg})^2}{N-1} \right]^{\frac{1}{2}},$$

$$(4)$$

where $\sigma$ is the standard deviation and N is the sample size. The k value is estimated with the coefficient of skewness assumes as zero [7].

Gumbel likelihood dissemination is broadly utilised for extraordinary worth examination of hydrologic and meteorological information, for example, floods, most extreme rainfalls and different occasions. The Gumbel formula used for the analysis is given in eqn (5):

$$X_T = X_{av} - \left( \frac{\sqrt{6}}{\pi} \right) \sigma \left[ 0.57721 + \ln \left\{ \ln \frac{T}{T-1} \right\} \right].$$

$$(5)$$

The above empirical relation holds great when the record length is 100 years or more [7].

### 3.2.2 Areal reduction and temporal pattern

Areal reduction factors (ARF's) and temporal pattern have not been published for storm event in Sarawak. Areal adjustments need to be made to the above point rainfalls to convert them to average catchment rainfall estimates. Because of the limited study on storm rainfall variation within Sarawak, ARF values based on a technical report by the US Weather Bureau (USWB) for application with the derived point rainfall estimates are adopted [8]. The ARF is available for various storm durations up to 72 hours and catchment area up to 1000 km$^2$. ARF was extrapolated for a longer duration and larger catchment to adopt in this study.

## 4  RESULT AND DISCUSSION

### 4.1  Rainfall frequency analysis

Design rainfall for various return period is estimated through rainfall frequency analysis using the 44 years historical rainfall record of the river basin. Rainfall frequency analysis was conducted using the Log Pearson Type III, Log-Normal and Gumbel method and the design rainfall for 2 to 100-year return period are adopted based on the highest value among the three methods. Table 1 shows the peak rainfall for 1-day and 3-day storm event for the 2-year, 5-year, 10-year, 50-year and 100-year return periods. Based on the analysis, for the 1-day storm, the 2-year and 100-year return period design rainfall are 115 mm and 206 mm, respectively. For the 3-day storm, the 2-year and 100-year return period design rainfall are 188 mm and 344 mm, respectively.

Table 1:  Design rainfall for 1-day and 3-day storm.

| Return period (year) | 1-day rainfall (mm) | 3-day rainfall (mm) |
|---|---|---|
| 2 | 115 | 188 |
| 5 | 138 | 228 |
| 10 | 155 | 256 |
| 50 | 191 | 318 |
| 100 | 206 | 344 |

4.2  Frequent peak flood analysis

After assessing the design rainfall for 2-year to 100-year return period, the design rainfalls were simulated through the RORB hydrological model to generate the peak flood discharge again each design rainfall.

Table 2 shows the peak flood discharge for 1-day and 3-day storm event for the 2-year, 5-year, 10-year, 50-year and 100-year return periods. Based on the analysis, the peak flood discharge for 1-days storms is higher than the 3-day storms. For the 1-day storm, the 2-year and 100-year return period peak, flood discharges are 3,867 $m^3/s$ and 7,043 $m^3/s$, respectively. For the 3-day storm, the 2-year and 100-year return period peak, flood discharges are 3,632 $m^3/s$ and 6,722 $m^3/s$, respectively.

Table 2:  Summary of the peak flood discharge for the various return period.

| Return period (year) | 1-day storm | 3-day storm |
|---|---|---|
| | Peak flood discharge ($m^3/s$) | Peak flood discharge ($m^3/s$) |
| 2 | 3,867 | 3,632 |
| 5 | 4,687 | 4,425 |
| 10 | 5,259 | 4,979 |
| 50 | 6,516 | 6,207 |
| 100 | 7,043 | 6,722 |

The values for 1-day and 3-day have been compared and assessed in terms of their variability; even the 1-day peak flood discharges values are higher than the 3-day peak flood discharges, both 1-day and 3-day peak flood discharges will be adopted to assess the flood risk assessment in Upper RRB.

## 5  CONCLUSIONS AND RECOMMENDATION
Based on the results of this study, it is noted that the peak flood discharge for 1-day storm events is higher than 3-day storm events. It would be interesting to assess the flood risk assessment under the 1-day, and 3-day peak flood discharges to assess the flood impact downstream of the Bakun Dam.

It is recommended to extend the analysis to derive the probable maximum precipitation (PMP) and probable maximum flood (PMF) to assess the flood impact under the extreme case.

## ACKNOWLEDGEMENTS
This study is funded by Sarawak Energy Berhad, Malaysia (SEB) under a Master's research program. The authors would like to thanks to SEB for the approval of a research grant and to allow using the data for hydro dams for this study. The authors also thankful to the Department of Irrigation and Drainage, Sarawak for providing the relevant rainfall data for the Rajang River Basin.

## REFERENCES
[1] Hussain, M., Yusof, K.W., Mustafa, M.R.U., Mahmood, R. & Jia, S., Evaluation of CMIP5 models for projection of future precipitation change in Bornean tropical rainforests. *Theoretical and Applied Climatology*, 2017.

[2]  Kourgialas, N. & Karatzas, G., Flood management and a GIS modelling method to assess flood-hazard areas-a case study. *Hydrological Sciences Journal*, **56**(2), pp. 212–225, 2011.
[3]  Melanie, L., Rahman, A. & Hagare, D., *An Investigation into Design Inputs for Design Flood Estimation in New South Wales*, 2012.
[4]  DID, *Sarawak Hydrological Year Book 2010*, Department of Irrigation and Drainage, Sarawak, 2016.
[5]  BM Alliance Coal Operations Pty Ltd., Flood hydrology technical report, Red Hill Mining Lease EIS, 2013.
[6]  Laurenson, R.G.M.E.M. & Nathan, R.J., *RORB Version 6: Runoff Routing Program User Manual*, Monash University and Hydrology and Risk Consulting Pty Ltd, Jan. 2010.
[7]  Sabarish, R.M., Narasimhan, R., Chandhru, A.R., Suribabu, C.R., Sudharsan, J. & Nithiyanantham, S., Probability analysis for consecutive-day maximum rainfall for Tiruchirapalli City (south India, Asia). *Applied Water Science,* **7**, pp. 1033–1042, 2017.
[8]  SMEC, Baleh hydroelectric project: Hydrology review and update, Sarawak Energy Berhad, Jul. 2013.

# SECTION 4
# FLOOD DAMAGE
# ASSESSMENT

# PERFORMANCES AND UNCERTAINTY OF TEMPERATURE METHODS FOR ILLICIT INFILTRATIONS AND INFLOWS ASSESSMENT IN STORMWATER SEWERS

UMBERTO SANFILIPPO, ANITA RAIMONDI, MARIANA MARCHIONI & GIANFRANCO BECCIU
Dipartimento di Ingegneria Civile e Ambientale (DICA), Politecnico di Milano, Italy

## ABSTRACT
The term "illicit flows" refers to all those unexpected and unwanted waters that are drained by or discharged into the urban drainage systems. Unmanaged illicit flows may cause significant losses in functionality both to the sewer networks and to the wastewater treatment plants. This paper focuses on one of the main approaches for illicit flow individuation and estimation in sewers that is based on the joint use of flow probes and temperature sensors; such an approach is often chosen for practical aims, because of its simplicity, affordability and adaptability. In particular, this paper is meant to fulfill the still existing lack of a systematic method to assess a priori its performances in terms of reliability and accuracy, by rigorously implementing the general uncertainty analysis theory. Then, through some meaningful numerical case studies, it is shown how to minimize the uncertainty on the illicit flow estimation when just one flow probe is available for the field survey. But it also highlights the potential weakness of the field results when the illicit flow rates are too small and/or their temperatures are too close in comparison to the regular flow. This research has been carried out within the framework of the project "PerFORM WATER 2030", funded by Regione Lombardia.
*Keywords: sewer, illicit flow, infiltration and inflow, flow probe, temperature sensor, uncertainty.*

## 1 INTRODUCTION
Sewer systems are extensive and aging structures, subject to cracks and misconnections during their operational lifetime. Illicit flows in sewer systems can be distinguished, depending on their source, into infiltrations and illicit inflows. The first are generally waters coming from aquifers or surface channels that can enter the network through cracked and broken pipes, leaky connections, or deteriorated manholes. The second are unauthorized and/or unintended connections delivering continuous or intermittent discharges to the sewer system. The temporal pattern of illicit flows can be constant or variable either in a random way or according to a regular time schedule, depending on their origin.

The presence of illicit flows in a sewer system causes lots of problems: network overload; increase of overflows; alterations of pumping systems functioning with arising energy costs and potential damages, especially if unexpected solid particles are transported; decrease of the efficiency of treatment plants due to flow dilution; and, last but not least, health and environmental problems.

Many regulations all over the world stress the importance to limit illicit flows into the stormwater sewers; it is very important not only to detect the so-called Infiltration & Inflow (I&I) into the drainage networks but also to understand their sources, trying to locate and quantify them. This knowledge is fundamental for the estimation of peak runoff [1] and the proper size of stormwater detention facilities [2]–[6] and Sustainable Urban Drainage Systems (SUDS) as like permeable pavements [7] and rainwater harvesting systems [8]–[10].

To this aim, the first step is usually the identification of the most critical areas and network sections, where further investigations must be developed at a local scale. The first step consist, most of the times, in the analysis of the night-time minimum of dry weather flows; after that, a number of different methodologies and technologies can be applied for the next

WIT Transactions on The Built Environment, Vol 194, © 2020 WIT Press
www.witpress.com, ISSN 1743-3509 (on-line)
doi:10.2495/FRIAR200101

level of the survey, depending on site conditions and resources availability: visual inspections and progressive sampling at manholes [11], smoke test and dye test [12], closed-circuit television camera (CCTV) inspections, Infra-Red camera [13], stable isotopes, polluting flows analysis, punctual measures of temperatures. In last decades, Distributed Time Sensing (DTS) technique has been successfully applied to locate I&I in sewers [14], [15]; it is based, too, on temperature measurements, but continuous in time and space; it allows to identify illicit flows for large distances without requiring access to private properties. Each one of these methodologies present some drawbacks: DTS requires cable installation into the sewer, dye and smoke tests are time consuming, visual inspections are – by definition – sensitive to human subjectivity, while methods based on sample analysis may be quite costly [16]; sometimes, a solution to improve the cost-effectiveness of a survey turns out to be the joint use of different technologies [17]. The method described in this paper needs punctual measures of temperature and flow rate, and can be performed indifferently by means of either DTS or simply temperature sensors coupled with flow probes: temperature anomalies, in fact, are generally indicators of illicit flows such as infiltration and inflow. The focus, here, is on the key aspect of the estimation of uncertainty about infiltrations or inflow rates calculated from the values of temperatures and in-sewer flow rates measured in the field, given the uncertainty of the temperature sensors and of the flow probes [18]. Then, the analysis of two meaningful numerical examples, described for each one of the three different general cases of interest for practical applications in real sewer systems, allows to evaluate the performances, in terms of accuracy and cost-benefits, which can be achieved by means of a given available set of temperature sensors and flow probes.

## 2 UNCERTAINTY ESTIMATION

Fig. 1 shows the reference scheme used in the following analysis.

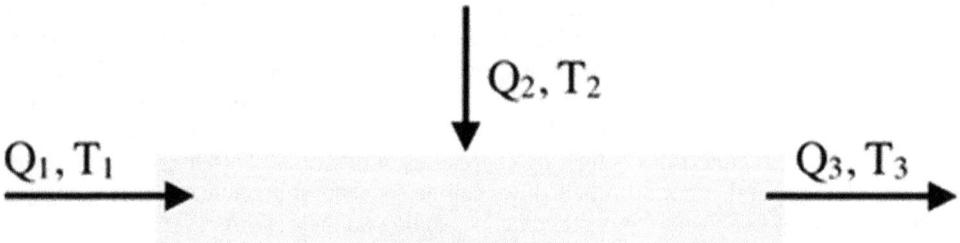

Figure 1: Scheme of reference for the analysis.

A generic portion of a sewer with undefined length and diameter can be considered as the control volume: inflow and outflow rates are respectively located in Sections 1 and 3 while in Section 2 an illicit entering flow rate is assumed (it can be generated from illicit connections, groundwaters or rainwater infiltrations). The six quantities of interest are the temperatures $T_1$, $T_2$ and $T_3$ and the flow rates $Q_1$, $Q_2$ and $Q_3$. Under the hypothesis of steady conditions, exploiting flow and energy conservation eqns (1), two of these six quantities can be calculated, if the remaining four are measured in the field:

$$\begin{cases} Q_1 + Q_2 = Q_3, \\ Q_1 \cdot T_1 + Q_2 \cdot T_2 = Q_3 \cdot T_3. \end{cases} \tag{1}$$

In the following analysis, three different cases are considered:

- Case A, which is typically related to infiltration of ground/rain waters: neither temperature nor flow rate of the illicit flow ($Q_2$ and $T_2$) can be measured; they are the unknown variables.
- Case B, which is typically related to illicit inflows: its temperature $T_2$ can be measured, while unknown variables are $Q_2$ and $Q_1$.
- Case C, which is typically related to illicit inflows, as well; its temperature $T_2$ can be measured, while unknown variables are $Q_2$ and $Q_3$.

Indeed, sometimes real situations can be quite different. For example, about case A, it may happen that a sensor for the measure of groundwater temperature can be located very close to the sewer, so that $T_2$ can be actually known; on the other hand, about case B and case C, due to the nature of the connection (e.g. a private property), it's not always easy to reach this punctual inflow and to measure its temperature $T_2$.

In general, if the measures of all the temperatures are available, for example when DTS is used, eqn (1) allows the estimation of the ratio between illicit flow rate $Q_2$ and one of the other flow rates in the pipe $Q_1$ or $Q_3$ [19]; but, of course, the two unknown flow rates can be both estimated only if one of the three flow rates is measured [20]. Discharge $Q_2$ is anyway the most important variable of the problem, by definition, as its knowledge drives the decisions about which kind of intervention is required to reduce the discovered illicit flow.

The computational procedure for uncertainty analysis, developed here for all the considered cases, allows to evaluate the performances which can be achieved, in terms of accuracy, in each one of the three above mentioned typical cases. In addition, this computational procedure is also the perfect tool to investigate and to compare the three cases, through some meaningful numerical examples, from the point of view of the reliability of the illicit flow estimations which can be obtained by field surveys. Based on such a comparison, it comes out which are, respectively, the most and the less favourable positioning schemes for temperature sensors and flow probes. It is assumed in each example that all the flow meters have the same relative uncertainty (expressed as %) while all the temperature sensors have the same absolute uncertainty (expressed as °C). By the way, the uncertainty about flow rates can also keep into account small short-term oscillations and variations and also transient storage effects [21], while the uncertainty about the temperatures can keep into account also the variations due to the thermic interactions between, on the one hand, the stream and, on the other hand, the channel walls and the air above the stream surface.

## 2.1  Case A

Unknown variables are illicit flow $Q_2$ and its temperature $T_2$; they can be estimated rearranging flow and energy conservation eqn (1):

$$Q_2 = Q_3 - Q_1, \qquad (2)$$

$$T_2 = \frac{Q_3 \cdot T_3 - Q_1 \cdot T_1}{Q_2}. \qquad (3)$$

Eqn (3) can be made dimensionless and written as:

$$\frac{T_2}{T_3} = \frac{Q_3}{Q_3 - Q_1} - \frac{Q_1}{Q_3 - Q_1} \cdot \frac{T_1}{T_3}. \qquad (4)$$

The relative uncertainty of illicit flow results:

$$\frac{U(Q_2)}{Q_2} = \left[\frac{U(Q)}{Q}\right]_{1;3} \cdot \alpha, \qquad (5)$$

with $\alpha = \sqrt{2 \cdot \left(\frac{Q_1}{Q_2}\right)^2 + 2 \cdot \left(\frac{Q_1}{Q_2}\right) + 1}$ and $\left[\frac{U(Q)}{Q}\right]_{1;3}$ is the relative uncertainty of the flow rate measurements, having assumed the same value for both the two measured flow rates $Q_1$ and $Q_3$. The relative uncertainty of estimated temperature $T_2$ results:

$$\frac{U(T_2)}{T_2} = \sqrt{\left[\frac{U(Q)}{Q}\right]_{1;3}^2 \cdot \left(\frac{(T_1 \cdot Q_1)^2 + (T_3 \cdot Q_3)^2}{(Q_2 \cdot T_2)^2} + \alpha^2\right) + [U^2(T)]_{1;3} \cdot \frac{1}{T_2^2} \cdot \alpha^2}. \qquad (6)$$

In particular, $[U^2(T)]_{1;3}$ is the uncertainty of temperature measurements, having assumed that it has the same value for both the two measured temperatures $T_1$ and $T_3$.

## 2.2  Case B

Unknown variables are the incoming flow rate $Q_1$ and the illicit flows $Q_2$, while all the temperatures and the downstream flow rate $Q_3$ are measured. Flow rates $Q_1$ and $Q_2$ can be estimated by the following eqns:

$$Q_2 = Q_3 \cdot \frac{\Delta T_{31}}{\Delta T_{21}}, \qquad (7)$$

with $\Delta T_{31} = T_3 - T_1$ and $\Delta T_{21} = T_2 - T_1$.
Once the discharge $Q_2$ has been obtained, $Q_1$ can be evaluated from the flow conservation eqn (1):

$$Q_1 = Q_3 - Q_2 = Q_3 \cdot \left(1 - \frac{\Delta T_{31}}{\Delta T_{21}}\right). \qquad (8)$$

The dimensionless expressions of eqns (7) and (8) result, respectively:

$$\frac{Q_2}{Q_3} = \frac{\Delta T_{31}}{\Delta T_{21}}, \qquad (9)$$

$$\frac{Q_1}{Q_3} = 1 - \frac{\Delta T_{31}}{\Delta T_{21}} = \frac{\Delta T_{23}}{\Delta T_{21}}. \qquad (10)$$

Relative uncertainties of estimated flow rates $Q_2$ and $Q_1$ result, respectively:

$$\frac{U(Q_2)}{Q_2} = \sqrt{\left[\frac{U(Q)}{Q}\right]_3^2 + [U^2(T)]_{1;2;3} \cdot \beta}, \qquad (11)$$

$$\frac{U(Q_1)}{Q_1} = \sqrt{\left[\frac{U(Q)}{Q}\right]_3^2 \cdot \left(\left(\frac{Q_2}{Q_1}\right)^2 + \left(\frac{Q_3}{Q_1}\right)^2\right) + \left(\frac{Q_2}{Q_1}\right)^2 \cdot [U^2(T)]_{1;2;3} \cdot \beta}, \qquad (12)$$

with $\beta = \frac{1}{\Delta T_{31}^2} \cdot \left(1 + \left(\frac{\Delta T_{32}}{\Delta T_{21}}\right)^2 + \left(\frac{\Delta T_{31}}{\Delta T_{21}}\right)^2\right)$ and $\Delta T_{32} = T_3 - T_2$.
In particular, $[U^2(T)]_{1;2;3}$ is the uncertainty on temperature measurements, having assumed that it has the same value for all of the three measured temperatures $T_1$, $T_2$ and $T_3$, while $\left[\frac{U(Q)}{Q}\right]_3$ is the relative uncertainty of the measured flow rate $Q_3$.

## 2.3  Case C

Also Case C assumes the knowledge of all temperatures $T_1$, $T_2$ and $T_3$, but it differs from Case B because it considers $Q_3$ instead of $Q_1$ as unknown variable. Resulting equations to estimate $Q_2$ and $Q_3$ can be obtained rearranging eqns (1):

$$Q_2 = Q_1 \cdot \frac{\Delta T_{31}}{\Delta T_{23}}, \tag{13}$$

$$Q_3 = Q_1 + Q_2 = Q_1 \cdot \left(1 + \frac{\Delta T_{31}}{\Delta T_{23}}\right), \tag{14}$$

with: $\Delta T_{23} = T_2 - T_3$.

Dimensionless expressions of eqns (13) and (14) result, respectively:

$$\frac{Q_2}{Q_1} = \frac{\Delta T_{31}}{\Delta T_{23}}, \tag{15}$$

$$\frac{Q_3}{Q_1} = 1 + \frac{\Delta T_{31}}{\Delta T_{23}} = \frac{\Delta T_{21}}{\Delta T_{23}}. \tag{16}$$

Relative uncertainties on estimated flow rates $Q_3$ and $Q_2$ result, respectively:

$$\frac{U(Q_2)}{Q_2} = \sqrt{\left[\frac{U(Q)}{Q}\right]_1^2 + [U^2(T)]_{1;2;3} \cdot \gamma}, \tag{17}$$

$$\frac{U(Q_3)}{Q_3} = \sqrt{\left[\frac{U(Q)}{Q}\right]_1^2 \cdot \left(\left(\frac{Q_1}{Q_3}\right)^2 + \left(\frac{Q_2}{Q_3}\right)^2\right) + \left(\frac{Q_2}{Q_3}\right)^2 \cdot [U^2(T)]_{1;2;3} \cdot \gamma}, \tag{18}$$

with $\gamma = \left(\dfrac{1 + \left(\frac{\Delta T_{31}}{\Delta T_{23}}\right)^2 + \left(\frac{\Delta T_{21}}{\Delta T_{23}}\right)^2}{\Delta T_{31}^{\,2}}\right)$; $\left[\dfrac{U(Q)}{Q}\right]_1$ is the relative uncertainty of the measured flow rate $Q_1$.

### 3  BEST PERFORMANCES IN ILLICIT FLOWS ESTIMATION

A trivial observation is that, of course, the more the temperatures $T_1$, $T_2$ and $T_3$ are similar, the less it is possible to get a reliable estimation of the illicit flow $Q_2$ in Case B and in Case C. Apart from this, one of the main targets of this paper, as already said, is to find out which is the best positioning of temperature sensors and flow probes in real sewer systems, when their task is to estimate illicit flow rates. With reference to the cases here discussed, in other words, the aim is to find out which one between Case A, Case B and Case C is the best in terms of accuracy and reliability of the illicit flow estimation. To do this, relative uncertainty of the measures of illicit flow rate $Q_2$ and temperature $T_2$ must be compared for the three cases; in particular, two different sets of assumptions for measurement uncertainty of flow rates and temperatures are here considered, respectively $U(Q_i) = 5\%$, $U(T_i) = 0.5°C$ and $U(Q_i) = 10\%$, $U(T_i) = 1°C$. Two examples for each case are carried out. Examples 1 can be considered as the benchmark: incoming flow rate $Q_1$ and illicit flow $Q_2$ are equal. Instead, Examples 2 is aimed to model a situation in which $Q_2$ is significantly smaller than $Q_1$.

### 3.1.1  Example 1

At first, unknown variables are estimated from measured data through eqns (2) and (3) for Case A1, eqns (7) and (8) for Case B1 and eqns (13) and (14) for Case C1:

Case A1: $Q_1 = 50$ l/s, $Q_3 = 100$ l/s, $T_1 = 20°C$, $T_3 = 18°C$ → $Q_2 = 50$ l/s, $T_2 = 16°C$
Case B1: $Q_3 = 100$ l/s, $T_1 = 20°C$, $T_2 = 30°C$, $T_3 = 25°C$ → $Q_1 = 50$ l/s, $Q_2 = 50$ l/s
Case C1: $Q_1 = 50$ l/s, $T_1 = 20°C$, $T_2 = 30°C$, $T_3 = 25°C$ → $Q_2 = 50$ l/s, $Q_3 = 60$ l/s

Then, through eqns (5) and (6) and considering the uncertainty values assumed for the measured quantities, relative uncertainties for $Q_2$ and $T_2$ in the three cases A1, B1 and C1 have been estimated (Table 1).

Graphical results for Cases A1, B1 and C1 for the assumed uncertainty values for flow rates and temperatures equal to $U(Q_i) = 5\%$ and $U(T_i) = 0.5°C$ are shown in Figs 2–4.

Table 1:  Assumed and calculated uncertainties of illicit flow rate and temperature in Example 1.

| Case | Assumed uncertainties for flow probes and temperature sensors | Calculated relative uncertainty $U(Q_2)/Q_2$ | Calculated relative uncertainty $U(T_2)/T_2$ |
|------|---------------------------------------------------------------|--------------------------------------------|--------------------------------------------|
| A1 | $U(Q_i) = 5\%$; $U(T_i) = 0.5°C$ | 11% | 18% |
| A1 | $U(Q_i) = 10\%$; $U(T_i) = 1°C$ | 22% | 37% |
| B1 | $U(Q_i) = 5\%$; $U(T_i) = 0.5°C$ | 13% | 2% |
| B1 | $U(Q_i) = 10\%$; $U(T_i) = 0.5°C$ | 26% | 3% |
| C1 | $U(Q_i) = 5\%$; $U(T_i) = 0.5°C$ | 25% | 2% |
| C1 | $U(Q_i) = 10\%$; $U(T_i) = 1°C$ | 50% | 3% |

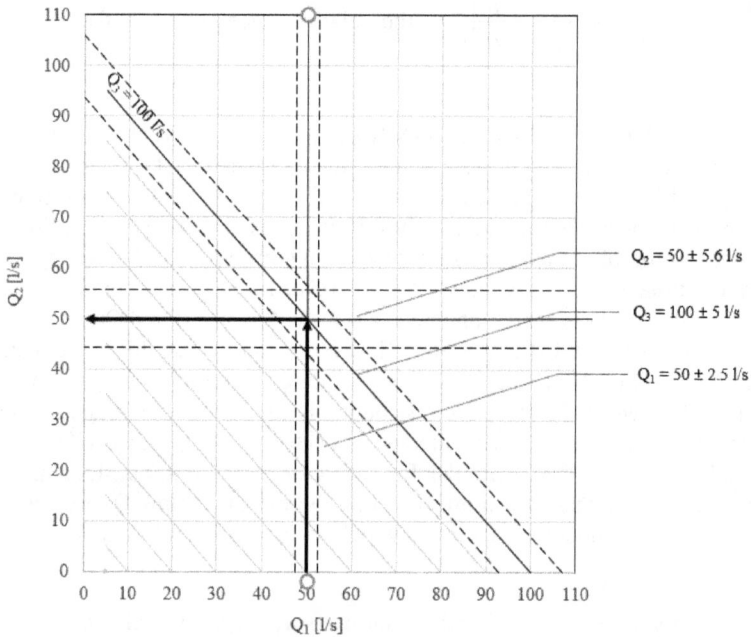

Figure 2:  Graphical representation of Case A1 with $U(Q_i) = 5\%$ and $U(T_i) = 0.5°C$.

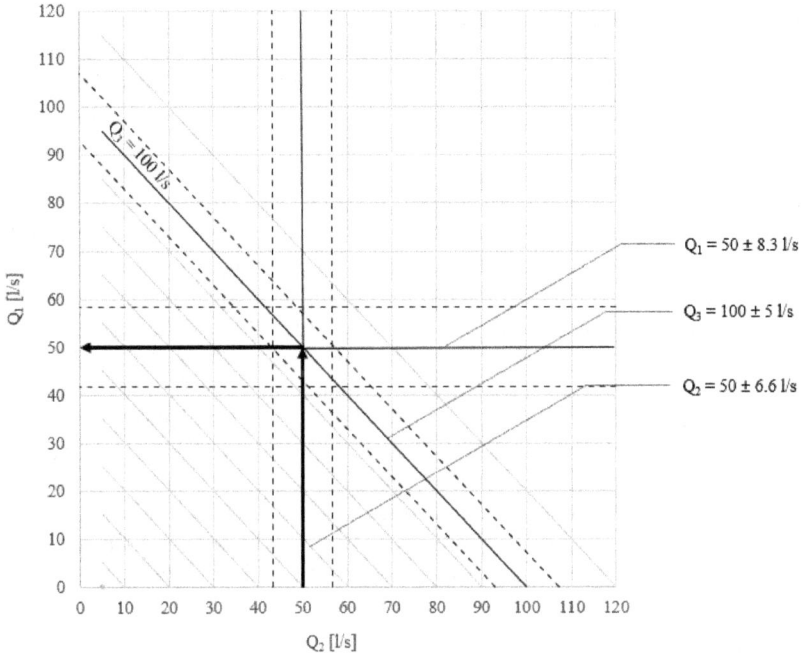

Figure 3: Graphical representation of Case B1 with $U(Q_i) = 5\%$ and $U(T_i) = 0.5°C$.

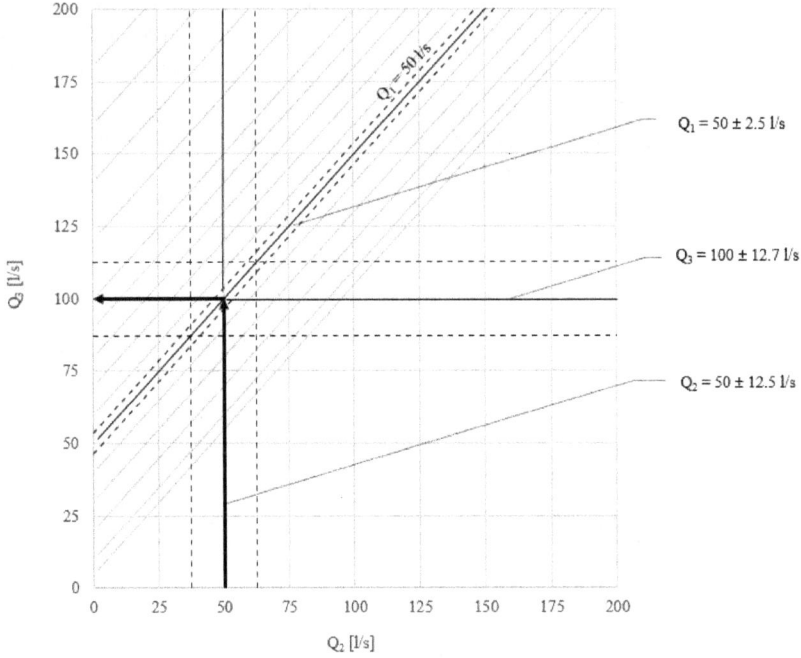

Figure 4: Graphical representation of Case C1 with $U(Q_i) = 5\%$ and $U(T_i) = 0.5°C$.

According to eqns (5), (11) and (18), a basic remark is that the values of the relative uncertainty $U(Q_i)/Q_i$ and of the absolute uncertainty $U(T_i)$ play the role of a multiplying coefficient in the expressions giving the relative uncertainty of the illicit flow $U(Q_2)$, in all of the three cases: so, in practice, doubling both the relative uncertainty of the measured flow rates $U(Q_i)/Q_i$ and the uncertainty of the measured temperature $U(T_i)$, the relative uncertainty $U(Q_2)/Q_2$ doubles. Beyond this, even more important is that it comes out that, fixed all the other parameters, the smallest relative uncertainty for illicit flow estimation $U(Q_2)/Q_2$ is for Case A1: this happens just because the sought flow rate $Q_2$ is calculated directly from the balance equation; differently, for cases B and C, flow rate $Q_2$ comes from energy conservation equation together with balance equation, with four measured quantities (and their own uncertainties) involved. But, on the other hand, flow probes are generally much more expensive than temperatures sensors; for this reason, Case B and Case C can be significantly less expensive than Case A, as they require just one flow probe instead of the two required in Case A. With reference to relative uncertainty of illicit flow temperature, Case A1 has the greatest uncertainty $U(T_2)$, since it is not measured directly in the field, while Case B1 and Case C1 have the smallest one, since $T_2$ temperatures are measured in the field in both of these cases. In conclusion, taking apart the operational aspects related to installation and running in each specific site, Case B1 could be seen as a kind of compromise in terms of cost-benefits, requiring a less expensive setup in comparison to Case A1 but having a better efficiency in comparison to Case C1.

### 3.1.2  Example 2

At first, unknown variables are estimated from measured data through eqns (2) and (3) for Case A1, eqns (7) and (8) for Case B1 and eqns (13) and (14) for Case C1:

Case A2: $Q_1 = 50$ l/s, $Q_3 = 60$ l/s, $T_1 = 20°C$, $T_3 = 18°C$ → $Q_2 = 10$ l/s, $T_2 = 8°C$
Case B2: $Q_3 = 100$ l/s, $T_1 = 23°C$, $T_2 = 30°C$, $T_3 = 25°C$ → $Q_1 = 71.4$ l/s, $Q_2 = 28.6$ l/s
Case C2: $Q_1 = 71.4$ l/s, $T_1 = 23°C$, $T_2 = 30°C$, $T_3 = 25°C$ → $Q_2 = 28.6$ l/s, $Q_3 = 100$ l/s

Then, through eqns (5) and (6) or simply starting from assumed uncertainties, relative uncertainties of $Q_2$ and $T_2$ in the three cases A2, B2 and C2 are estimated in Table 2.

Graphical results for Cases A2, B2 and C2, for assumed uncertainties of flow rates and temperatures equal to $U(Q_i) = 5\%$ and $U(T_i) = 0.5°C$, are shown in Figs 5–7.

Table 2:   Assumed and calculated uncertainties of illicit flow rate and temperature in Example 2.

| Case | Assumed uncertainties | Calculated relative uncertainty $U(Q_2)/Q_2$ | Calculated relative uncertainty $U(T_2)/T_2$ |
|---|---|---|---|
| A2 | $U(Q_i) = 5\%$; $U(T_i) = 0.5°C$ | 39% | 111% |
| A2 | $U(Q_i) = 10\%$; $U(T_i) = 1°C$ | 78% | 222% |
| B2 | $U(Q_i) = 5\%$; $U(T_i) = 0.5°C$ | 32% | 2% |
| B2 | $U(Q_i) = 10\%$; $U(T_i) = 0.5°C$ | 64% | 3% |
| C2 | $U(Q_i) = 5\%$; $U(T_i) = 0.5°C$ | 44% | 2% |
| C2 | $U(Q_i) = 10\%$; $U(T_i) = 1°C$ | 89% | 3% |

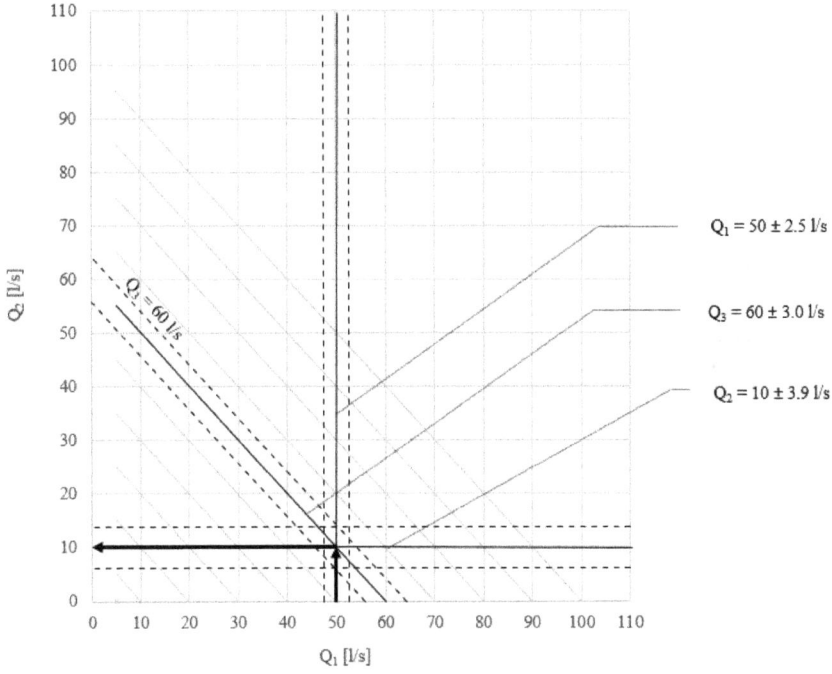

Figure 5: Graphical representation of Case A2 with $U(Q_i) = 5\%$ and $U(T_i) = 0.5°C$.

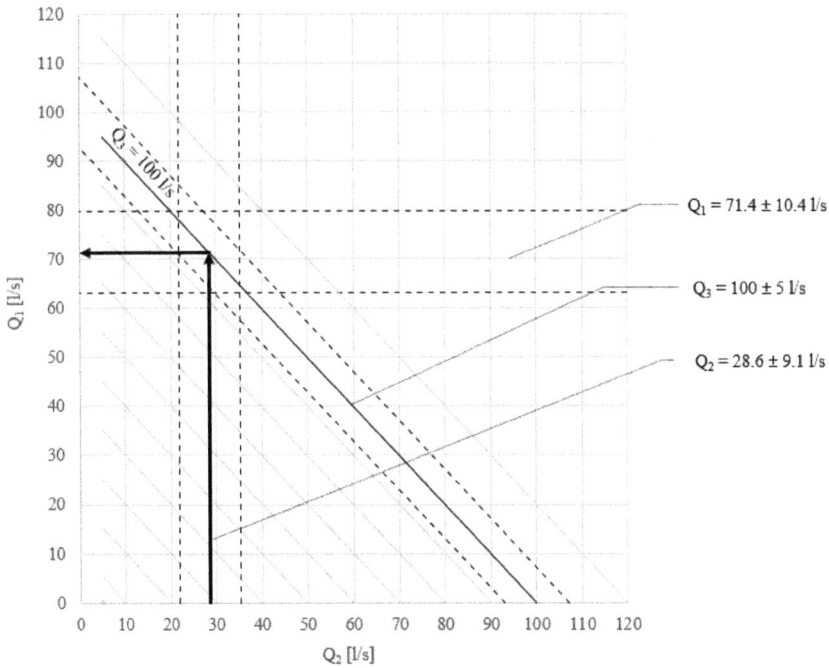

Figure 6: Graphical representation of Case B2 with $U(Q_i) = 5\%$ and $U(T_i) = 0.5°C$.

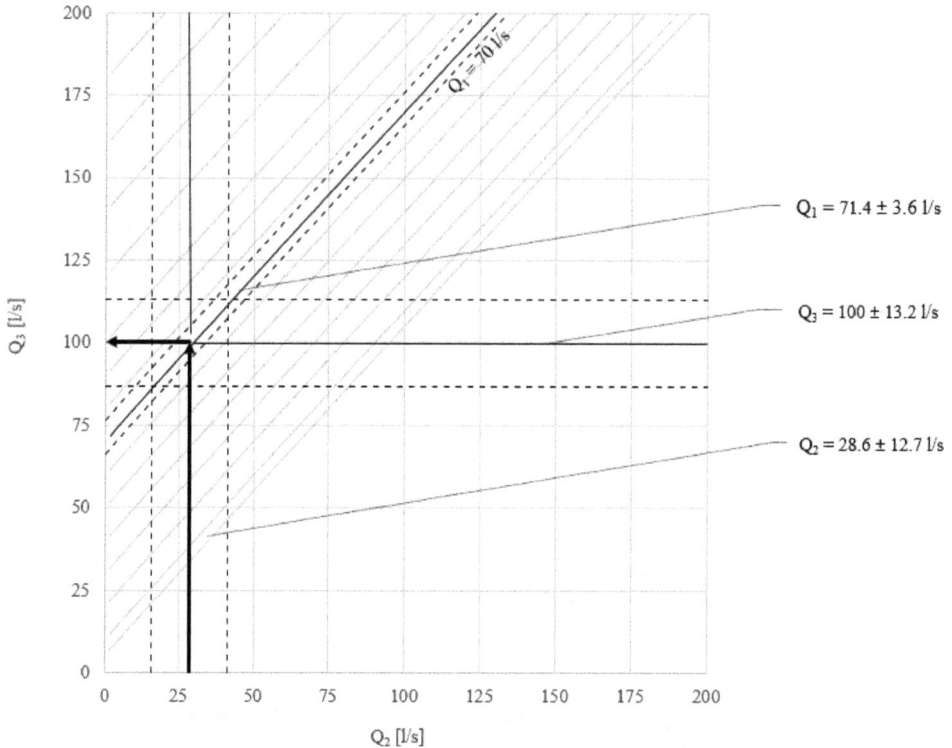

Figure 7:   Graphical representation of Case C2 with $U(Q_i) = 5\%$ and $U(T_i) = 0.5°C$.

According to Table 2, Case B2 turns out to be, although slightly, the most suitable, having the lowest relative uncertainty concerning both flow rates and temperatures. Another key outcome is that relative uncertainties of flow rates and temperatures are systematically much higher in comparison with those of Example 1 reported in Table 1; this means that, for a given case (no matter if Case A, Case B or Case C), when illicit flow $Q_2$ and upstream flow $Q_1$ are not too different then the resulting relative uncertainties are lower than what they are when $Q_2$ is quite (or much) smaller than $Q_1$.

## 4  CONCLUSIONS

This paper is meant to offer a support to operators involved in planning and executing field surveys for the estimation of illicit inflows in sewers in steady conditions. As a matter of fact, this paper has described the mathematical procedures required to estimate both illicit flows and their temperatures by means of a proper set of temperature sensors and flow probes. It has also provided a detailed description of the mathematical procedures to assess the uncertainty of these estimated illicit flow rates and temperatures.

In addition, the numerical examples developed for the three typical cases of practical interest has pointed out that the more the illicit flows and the sewer flow are different, the less it is possible to estimate them in a reliable way. It has also emerged that when just one single flow probe is available for the field survey and the suspected connection is not accessible, to minimize the uncertainty on the illicit flow estimation it is better to place the available flow probe upstream instead of downstream the sewer trunk under investigation.

## REFERENCES

[1]  Becciu, G. & Paoletti, A., Random characteristics of runoff coefficient in urban catchments. *Water Science and Technology*, **36**, pp. 39–44, 1997.

[2]  Becciu, G. & Raimondi, A., Probabilistic analysis of spills from stormwater detention facilities. *WIT Transactions on the Built Environment*, vol. 139, WIT Press: Southampton and Boston, pp. 159–170, 2014. DOI: 10.2495/UW140141.

[3]  Becciu, G. & Raimondi, A., Probabilistic analysis of the retention time in stormwater detention facilities. *Procedia Engineering*, **119**(1), pp. 1299–1307, 2015. DOI: 10.1016/j.proeng.2015.08.951.

[4]  Becciu, G. & Raimondi, A., Probabilistic modeling of the efficiency of a stormwater detention facility. *International Journal of Sustainable Development and Planning*, **10**(6), pp. 795–805, 2015. DOI: 10.2495/SDP-V10-N6-795-805.

[5]  Raimondi, A. & Becciu, G., On pre-filling probability of flood control detention facilities. *Urban Water Journal*, **12**(4), pp. 344–351, 2015. DOI: 10.1080/1573062X.2014.901398.

[6]  Raimondi, A. & Becciu, G., On the efficiency of stormwater detention tanks in pollutant removal. *International Journal of Sustainable Development and Planning*, **12**(1), pp. 144–154, 2017. DOI: 10.2495/SDP-V12-N1-144-154.

[7]  Marchioni, M. & Becciu, G., Experimental results on permeable pavements in urban areas: A synthetic review. *International Journal of Sustainable Development and Planning*, **10**(6), pp. 806–817, 2015.

[8]  Becciu, G., Raimondi, A. & Dresti, C., Semi-probabilistic design of rainwater tanks: A case study in Northern Italy. *Urban Water Journal*, **15**(3), pp. 192–199, 2018. DOI: 10.1080/1573062X.2016.1148177.

[9]  Raimondi, A. & Becciu, G., Probabilistic modeling of rainwater tanks. *Procedia Engineering*, **89**, pp. 1493–1499, 2014. DOI: 10.1016/j.proeng.2014.11.437.

[10]  Raimondi, A. & Becciu, G., Probabilistic design of multi-use rainwater tanks. *Procedia Engineering*, **70**, pp. 1391–1400, 2014. DOI: 10.1016/j.proeng.2014.02.154.

[11]  Butler, D. & Davies, J., *Urban Drainage*, CRC Press, 2004.

[12]  Hoes, O.A.C., Schilperoort, R.P.S., Luxemburg, W.M.J., Clemens, F.H.L.R. & Van de Giesen, N.C., Locating illicit connections in storm water sewers using fibre-optic distributed temperature sensing. *Water Research*, **43**, pp. 5187–5197, 2009.

[13]  Lepot, M., Makris, K.F. & Clemens, F.H.L.R., Detection and quantification of lateral, illicit connections and infiltration in sewers with Infra-Red camera: Conclusions after a wide experimental plan. *Water Research*, **122**, pp. 678–691, 2017.

[14]  Nienhuis, J., de Haan, C., Langeveld, J., Klootwijk, M. & Clemens, F.H.L.R., Assessment of detection limits of fiber-optic distributed temperature sensing for detection of illicit connections. *Water Science and Technology*, **67**(12), pp. 2712–2718, 2013.

[15]  Schilperoort, R.P.S., Gruber, G., Flamink, C.M.L., Clemens, F.H.L.R. & van der Graaf, J.H.M.H., Temperature and conductivity as control parameters for pollution-based real-time control. *Water Science and Technology*, **54**(11–12), pp. 257–263, 2006.

[16]  Panasiuk, O., Hedström, A., Marsalek, J., Ashley, R.M. & Viklander, M., Contamination of storm water by wastewater: A review of detection method. *Journal of Environmental Management*, **152**, pp. 241–250, 2015.

[17]  Beheshti, M. & Saegrov, S., Detection of extraneous water ingress into the sewer system using tandem methods: A case study in Trondheim city. *Water Science and Technology*, **79**, pp. 231–239, 2019.

[18] Lazzarin, A., Orsi, E. & Sanfilippo, U., A statistical analysis on experimental calibration data for flowmeters in pressure pipes. *Journal of Physics Conference Series*, **882**, 2017. DOI: 10.1088/1742-6596/882/1/012015.

[19] Beheshti, M. & Saegrov, S., Quantification assessment of extraneous water infiltration and inflow by analysis of the thermal behaviour of the sewer network. *Water,* **10**, pp. 1070–1087, 2018.

[20] Schilperoort, P.S. &. Clemens, F.H.L.R., Fibre-optic distributed temperature sensing in combined sewer systems. *Water Science and Technology*, **60**(5), pp. 1127–34, 2009.

[21] Schmid, B.H., Innocenti, I. & Sanfilippo, U., Characterizing solute transport with transient storage across a range of flow rates: The evidence of repeated tracer experiments in Austrian and Italian streams. *Advances in Water Resources*, **33**(11), pp. 1340–1346, 2010. DOI: 10.1016/j.advwatres.2010.06.001.

# FLOOD DAMAGE FUNCTIONS IN THE VRBAS RIVER BASIN

RADUŠKA CUPAĆ[1], EDIN ZAHIROVIĆ[2] & VUJADIN BLAGOJEVIĆ[3]
[1]United Nations Development Programme B&H, Bosnia and Herzegovina
[2]Centre for development and support Tuzla, Bosnia and Herzegovina
[3]Institute for water management Bijeljina, Bosnia and Herzegovina

## ABSTRACT

The primary task of this research is the development of flood damage functions in order to demonstrate the level of damage caused to household assets and agriculture by floods in the Vrbas river basin in Bosnia and Herzegovina. Damage functions describe the connection between the flood depth and the value of flood-affected assets and potential damage. In the process of developing the function, information on representative buildings viewed from the perspective of civil engineering and electrical machinery, as well as representative building with its movables, were used. Results of the damage functions relate to synthetic damage that is not the result of registered damage in the past. It consists of the following steps: (1) selection of representative residential buildings and an assessment of their market value and the value of any movables, (2) determination of the characteristics of flood levels above ground level through hydrology and the use of hydraulic models, (3) expert assessment of the resulting damage in accordance with the specific flood levels in those buildings, and (4) determination of the value of the damage caused to the buildings taking into consideration the increase in damage in relation to the flood depth. The total agricultural flood damage in the basin was assessed as the result of the sum of losses in the floodplain based on the most represented crops. The total cost of the flood damage was obtained by multiplying the losses incurred because of the reduction in yields expressed in Euro/ha and based on the surface area of crops in the floodplain (ha). Damage functions were used to calculate the potential flood damage in the Vrbas river basin. The potential amount of damages calculated for households and agriculture was EUR 47 mil. It was in accordance with recorded actual flood damages from 2014 in the Vrbas river basin, so they can be used for future analyses.

*Keywords: damage function, flood depth, potential damage, Vrbas river basin, household assets.*

## 1 INTRODUCTION

Damage functions represent the relationship between the flood depth and the value of flood affected assets and potential damage. In the relevant literature, the main parameter for determining the damage value is the flood depth or more precisely the flood depth in the building. In addition to flood depth, the water speed of the flood wave, the duration of the flood, the seasonality or time of the flooding affect the amount of damage; however, these parameters are rarely included in the calculation of the amount of damage [1]. Two main approaches are distinguishable in the development of flood damage models: empirical approaches that use damage data collected after flood events and synthetic approaches that use damage data collected via "what if" questions. "What if" analyses estimate the level of damage expected in the case of a certain flood situations, namely "Which damage would you expect if the flood depth was 2 m above the building floor?" [2]. Penning-Rowsell et al. present examples for this approach [3]. Results of the damage functions addressed in this analysis are synthetic. They are not the result of registered damage in the past but the result of a process that comprised the following steps:

1. selection of representative residential buildings and an assessment of their value as well as the value of any movables in those buildings;
2. determination of the characteristics of flood levels above ground level through hydrology and the use of hydraulic models;

WIT Transactions on The Built Environment, Vol 194, © 2020 WIT Press
www.witpress.com, ISSN 1743-3509 (on-line)
doi:10.2495/FRIAR200111

3.  expert assessment of the resulting damage caused to the representative buildings in accordance with the specific flood levels in those buildings;
4.  determination of the value of the damage caused to the buildings taking into consideration the increase in damage in relation to the flood depth, in fact determining the damage function.

The synthetic damage functions addressed in this analysis are for projection purposes and relate mainly to future and consequent calculations in cost-benefit analysis and the analysis of asset insurance and the flood protection system. To select the representative residential buildings and estimate their value, telephone interviews were conducted covering all settlements within the Vrbas river basin exposed to flood risk. The Vrbas river basin is part of the Danube river basin. It is located in north western Bosnia and Herzegovina (BH) and covers an area of 6,386 km². It is not usual to use site specific flood characteristics for developing the damage functions, but the reason for selecting only Vrbas river basin when creating the damage functions is the existence of relevant hydrology, hydraulic, socio-economic and other flood-related data for this river basin. For other basins in BH the data are not specified, particularly hydrology and hydraulic data which do not exist or their making is ongoing. All these data were generated during the project of flood risk mitigation in the Vrbas river basin implemented by the United Nations Development Programme (UNDP). The average value of the household assets was obtained through a phone assessment that covered 3,500 households more or less exposed to flood risk utilising a stratified random sample that took into consideration representation per the criteria urban and rural. The settlements were selected for strata and the average number of respondents per settlement was 5% of the total population in a given settlement. In addition to general questions, the interview included questions on residential and support buildings and their features.

Two types of data are used to calculate the flood damage curve for households (according to the principle of % assets value). The first, average value of household assets and the second related to the average damage on building and movables. Putting in ratio the damage on building (with or without movables) and average market value of household assets (with or without movables), the percentage of damage for household assets (with or without movables) is obtained for different flood levels. To calculate the first data, the average values of household assets, market value is used which takes into consideration the material of which the buildings are made, number of floors in building, location of building, connection of building with utility infrastructure, dimensions, year of construction and terms of building maintenance including the value of movables. Use of the market value of household assets reflects the actual differences in development level of certain municipalities within the assessed region. The following market values were considered in order to calculate the value of the household assets: the value of apartment/house, fence, heating system as well as significant electricity users, significant furniture, agricultural equipment, fodder stocks and livestock. This combined with the value of individual assets is how the average value of the assets per household per settlement is presented. To calculate the second data, the amount of damage on building and movables, the costs of bringing the building and associated movables into the initial condition are taken into consideration, especially rebuilding and replacement costs.

Flood hazard and flood risk maps developed for the Vrbas river basin were used to determine values in relation to the characteristic flood levels. The expert assessment of risk to representative buildings in line with the characteristic flood levels was conducted at representative residential buildings in the Vrbas river basin. The damage to these buildings was assessed in terms of civil engineering and electrical machinery based on flood depth water levels in buildings set at 0.5 m, namely from 0.5–5 m. Depending on the water depth in a building, the cost of the works required to return the building to function or its initial condition, in fact the rebuilding costs, were assessed.

In relation to the civil engineering aspect, the following types of work were considered: repair of outside fences, cleaning the well and repairing the hydro pack, emptying the septic tanks, drying the walls, removal of the existing internal and exterior joinery and the installation of new joinery, painting the walls and ceiling, removal of the existing wooden floor (parquet, ship floor) and the purchase and installation of new floor.

In relation to the electrical machinery aspect, the following types of works were considered: purchase of materials and electrical installation works; purchase of all materials for the installation of the main electrical box with all necessary measures of electrical protection and the space for the installation of the meter and counter and installation of the appropriate set including all necessary works as well as grounding works; purchase and installation of a single-phase two-tariff meter; purchase of the materials and balancing the potential for water installations; obtain the certificate and registration for electrical installation testing in house; repair the heating boiler; replacement of the main electronic board and internal ventilator in boilers and/or cleaning and replacement of damaged parts in individual stoves.

The cost of the reconstruction works required to repair the damage caused to a building was determined taking into consideration all of the above-mentioned types of works for all flood levels in buildings ranging from 0.5–5 m. In this way, a unified damage function pertaining to representative buildings in the Vrbas river basin was established. The usual practice is to present the damage as a percentage of the total value of the building [4], [5]. This was done by comparing the average value of representative buildings in the Vrbas river basin in accordance with the equation below

$$\% \, damage = \frac{reconstruction \; costs}{total \; value \; of \; representative \; building}.$$

To determine the damage function in a representative building with movables, in addition to the reconstruction costs, the cost of replacing significant electrical appliances, furniture as well as agricultural equipment and other movables are included. In this case, the damage is presented as a percentage of the value of the total household assets in accordance with the equation below

$$\% \, damage = \frac{reconstruction \; costs + replacement \; of \; movables}{total \; value \; of \; household \; assets}.$$

Since the average value of assets in household was determined for each of the 13 municipalities in the Vrbas river basin, 13 damage functions were established based on the value of the assets. All 13 damage functions were analysed per municipality leading to the conclusion that it is possible to establish three similar groups of municipalities to be presented with three damage functions. Grouping the municipalities in order to determine damage functions was done according to the household assets value criteria per municipalities/cities except for Banja Luka city. One damage function relates to the City of Banja Luka as it is business, administrative, academic and cultural centre and exceptionally urban area with over 60% of total population from the Vrbas river basin. Another damage function relates to the average value of household assets over 100,000 Bosnia and Herzegovina convertible mark – BAM (EUR 51,129) for the municipalities of Gradiška (156,035 BAM), Jajce (101,101 BAM), Kotor Varoš (128,895 BAM), Laktaši (124,236 BAM) and Srbac (119,434 BAM). Average value of household assets to 100,000 BAM (EUR

51,129) includes the municipalities which make the third damage function, such as: Bugojno municipality (average value of household assets 92,144 BAM), Čelinac (90,652 BAM), Donji Vakuf (73,386 BAM), Gornji Vakuf (85,257 BAM), Jezero (86,787 BAM), Šipovo (78,077 BAM) and Mrkonjić Grad (75,376 BAM). The value of household assets reflects the development level of municipalities and cities in BH frames.

## 2 DAMAGE FUNCTION DEPENDING ON THE FLOOD DEPTH FOR A BUILDING

Damage function depends on the flood depth in a representative building and this was obtained for buildings without movables based on the average value of representative buildings in the Vrbas river basin and the reconstruction costs for all flood levels in buildings (from 0.5–5 m).

The average value of a representative building in the Vrbas river basin is BAM 94,415 (EUR 48,274) without movables. In line with the average value of representative buildings in the Vrbas river basin, the percentage value of damage obtained is as presented in Table 1 and Fig. 1.

Table 1: Reconstruction costs for representative buildings in the Vrbas river basin with the damage value set according to the flood level. *(Source: Author.)*

| Water depth in building (x) | Reconstruction costs | % damage value to building (x) (without movables) |
|---|---|---|
| 0.5 m | 7,924 BAM (4,051 EUR) | 8.4% |
| 1.0 m | 11,870 BAM (6,069 EUR) | 12.4% |
| 1.5 m | 12,618 BAM (6,451 EUR) | 13.2% |
| 2.0 m | 12,885 BAM (6,588 EUR) | 13.5% |
| 2.5 m | 13,204 BAM (6,751 EUR) | 13.8% |
| 3.0 m | 16,306 BAM (8,337 EUR) | 17.1% |
| 3.5 m | 16,339 BAM (8,354 EUR) | 17.1% |
| 4.0 m | 18,759 BAM (9,591 EUR) | 19.7% |
| 4.5 m | 18,824 BAM (9,625 EUR) | 19.7% |
| 5.0 m | 19,000 BAM (9,715 EUR) | 19.9% |

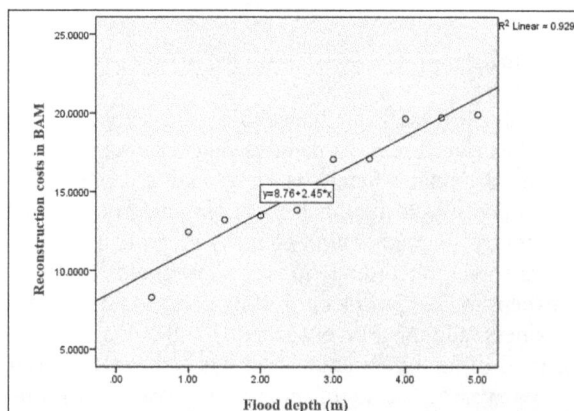

Figure 1: Damage function (curve) as a share of the assets value depending on the water depth in a representative building, excluding movables. *(Source: Author.)*

In order to conduct interpolation and/or extrapolation of any damage value against the flood depth it is necessary to present the damage function in relation to the flood depth in a building without movables. Therefore, depending on the flood depth in a building without movables, the damage function in the Vrbas river basin reads as shown below

$$y_{BP} = 8,760 + 2,450x, \tag{1}$$

where $y_{BP}$ represents the reconstruction costs in BAM (1 EUR equals 1.95583 BAM) and $x$ represents the flood depth, which is in fact the water depth in the building expressed in metres.

Presentation of the damage functions, utilizing linear function, was done due to the simplicity of the function, which also has high R-squared. As already mentioned, the data for damage functions were generated during implementation of the flood risk mitigation project in the Vrbas river basin for 13 municipalities/cities in this area. Those functions, among other things, are to be used by local and regional authorities addressing the flood risk mitigation issues. Every non-linear presentation of damage function could result in reluctance of authorities to use the damage functions and would return to non-selective and scientifically ungrounded flood risk mitigation approach.

## 3 DAMAGE FUNCTION DEPENDING ON THE FLOOD DEPTH FOR BUILDINGS WITH MOVABLES

Depending on the flood depth, the damage functions for representative buildings with movables was obtained based on the average value of assets in households per municipality within the Vrbas river basin, the reconstruction costs for all flood levels in buildings (from 0.5–5 m) as well as the costs replacing movables (significant electrical appliances, furniture, agricultural equipment and other movables).

As previously mentioned, it was concluded that it is possible to establish three groups of municipalities presented through three damage functions. The following text explains these damage functions.

### 3.1 Damage function depending on the flood depth for buildings with movables in the City of Banja Luka

The average value of assets in the representative households in the City Banja Luka amounted to BAM 145,768 (EUR 74,530), including the civil engineering and electrical machinery value of buildings with movables. Table 2 and Fig. 2 were obtained based on an analysis of the reconstruction costs for all flood levels in the representative buildings (from 0.5–5 m).

To conduct interpolation and/or extrapolation of any damage value (as a share of the assets value) against the flood depth it is necessary to present the damage function. Therefore, depending on the flood depth in a building in the City of Banja Luka, the damage function reads as shown below

$$y_{BL} = 1.6 \cdot x + 9.54, \tag{2}$$

where $y_{BL}$ represents the percentage of damage to the assets value for households in the City of Banja Luka and $x$ represents the flood depth (in fact the water depth in the building expressed in metres).

Table 2: Reconstruction costs for representative buildings and the replacement of movables within the city Banja Luka and the percentage value of damage for representative buildings. *(Source: Author.)*

| Water depth in building (x) | Cost of reconstruction and replacement of movables | % of damage to the assets value in households |
|---|---|---|
| 0.5 m | 13,479 BAM (6,892 EUR) | 9.25% |
| 1.0 m | 17,425 BAM (8,909 EUR) | 11.95% |
| 1.5 m | 18,173 BAM (9,292 EUR) | 12.47% |
| 2.0 m | 18,440 BAM (9,428 EUR) | 12.65% |
| 2.5 m | 18,759 BAM (9,591 EUR) | 12.87% |
| 3.0 m | 21,861 BAM (11,177 EUR) | 15.00% |
| 3.5 m | 21,894 BAM (11,194 EUR) | 15.02% |
| 4.0 m | 24,314 BAM (12,432 EUR) | 16.68% |
| 4.5 m | 24,379 BAM (12,465 EUR) | 16.72% |
| 5.0 m | 24,555 BAM (12,555 EUR) | 16.85% |

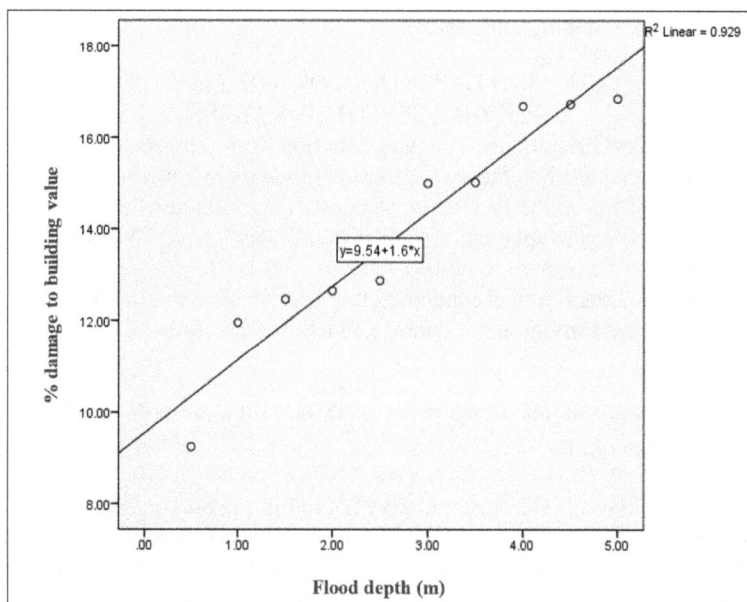

Figure 2: Damage function (curve) as a share of the assets value depending on the water depth in buildings in the city of Banja Luka. *(Source: Author.)*

## 3.2 Damage function depending on flood depth in building with movables in the municipalities of Gradiška, Jajce, Kotor Varoš, Laktaši and Srbac

The average value of assets in the households within the second group of municipalities totals 125,940 BAM (64,392 EUR). Based on the analysis of the reconstruction costs at all flooded levels of building (from 0.5–5 m) as well as estimated amount needed for replacement of movables affected by floods, Table 3 and Fig. 3 are obtained.

Table 3:   Reconstruction costs, replacement of movables and a percentage of the damage value for the second group of municipalities. *(Source: Author.)*

| Water depth in building (x) | Cost of reconstruction and the replacement of movables | % of damage to the assets value in households |
|---|---|---|
| 0.5 m | 15,243 BAM (7,794 EUR) | 12.10% |
| 1.0 m | 19,189 BAM (9,811 EUR) | 15.24% |
| 1.5 m | 19,937 BAM (10,194 EUR) | 15.83% |
| 2.0 m | 20,204 BAM (10,330 EUR) | 16.04% |
| 2.5 m | 20,523 BAM (10,493 EUR) | 16.30% |
| 3.0 m | 23,625 BAM (12,079 EUR) | 18.76% |
| 3.5 m | 23,658 BAM (12,096 EUR) | 18.79% |
| 4.0 m | 26,078 BAM (13,333 EUR) | 20.71% |
| 4.5 m | 26,143 BAM (13,367 EUR) | 20.76% |
| 5.0 m | 26,319 BAM (13,457 EUR) | 20.90% |

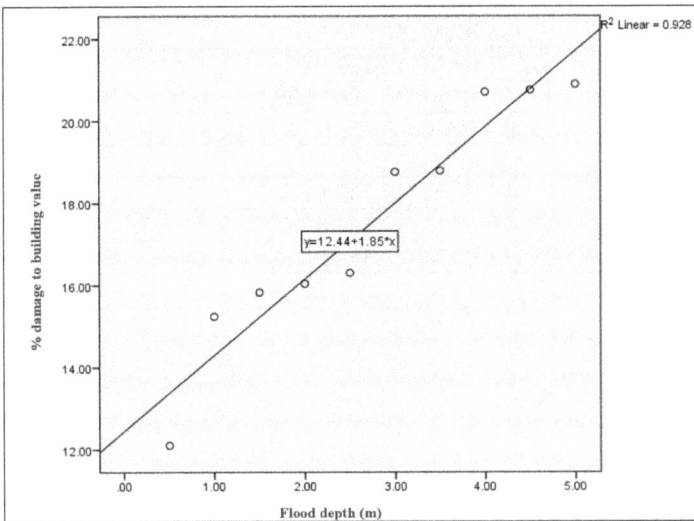

Figure 3:   Damage function (curve) as a share of the assets value depending on the water depth in buildings in the municipalities of Gradiška, Jajce, Kotor Varoš, Laktaši and Srbac. *(Source: Author.)*

To conduct interpolation and/or extrapolation of any damage value (as a share of the assets value) against the flood depth it is necessary to present the damage function. Therefore, the damage function, depending on the flood depth in buildings in the municipalities of Gradiška, Jajce, Kotor Varoš, Laktaši and Srbac reads as shown below

$$y_{DG} = 1.85 \cdot x + 12.44, \qquad (3)$$

where $y_{DG}$ represents the percentage of damage to the assets value in the municipalities of Gradiška, Jajce, Kotor Varoš, Laktaši and Srbac and $x$ represents the flood depth (in fact the water depth in the building expressed in metres).

### 3.3 Damage function depending on the flood depth in buildings with movables in the municipalities of Bugojno, Čelinac, Donji Vakuf, Gornji Vakuf, Jezero, Šipovo and Mrkonjić Grad

The average value of assets in the households within the third group of municipalities amounted to BAM 83,097 (42,487 EUR), including the civil engineering and electrical machinery value of buildings with movables. Table 4 and Fig. 4 are based on an analysis of the reconstruction costs for all flood levels in these buildings (from 0.5–5 m) as well as the estimated amount needed to replace movables affected by the floods.

Table 4: Reconstruction costs, replacement of movables and a percentage of the damage value for the third group of municipalities. *(Source: Author.)*

| Water depth in building (x) | Cost of reconstruction and the replacement of movables | % of damage to the assets value of households |
|---|---|---|
| 0.5 m | 15,040 BAM (7,690 EUR) | 18.10% |
| 1.0 m | 18,987 BAM (9,708 EUR) | 22.85% |
| 1.5 m | 19,735 BAM (10,090 EUR) | 23.75% |
| 2.0 m | 20,002 BAM (10,227 EUR) | 24.07% |
| 2.5 m | 20,321 BAM (10,390 EUR) | 24.45% |
| 3.0 m | 23,423 BAM (11976 EUR) | 28.19% |
| 3.5 m | 23,456 BAM (11,993 EUR) | 28.23% |
| 4.0 m | 25,876 BAM (13,230 EUR) | 31.14% |
| 4.5 m | 25,940 BAM (13,263 EUR) | 31.22% |
| 5.0 m | 26,117 BAM (13,353 EUR) | 31.43% |

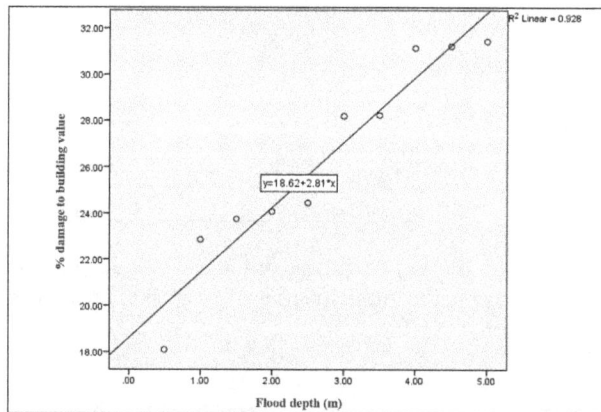

Figure 4: Damage function (curve) as a share of the assets value depending on the water depth in buildings for the third group of municipalities. *(Source: Author.)*

To conduct interpolation and/or extrapolation of any damage value (as a share of the assets value) against the flood depth it is necessary to present the damage function. Therefore, the damage function depending on the flood depth in a building in the municipalities of Bugojno, Čelinac, Donji Vakuf, Gornji Vakuf, Jezero, Šipovo and Mrkonjić Grad reads shown below

$$y_{TG} = 2.81x + 18.62, \qquad\qquad (4)$$

where $y_{TG}$ represents the percentage of damage to the assets value in the municipalities of Bugojno, Čelinac, Donji Vakuf, Gornji Vakuf, Jezero, Šipovo and Mrkonjić Grad and $x$ represents the flood depth (in fact the water depth in the building expressed in metres).

3.4  A comparison between damage functions in the Vrbas river basin and other regions

So far, there is no generally accepted methodology for the construction of these functions because they reflect local conditions and therefore vary from country to country. For comparison purposes the functions of potential damage to the rivers Rhine and Elba in Germany [6] and a the case study of the City of Palermo in Italy [7] and the Municipality of Moschato in Greece [8] as well as guidelines on damage functions for different types of residential buildings that was developed by the US Army Corps of Engineers [9] were considered.

After considering the damage functions for the above-mentioned research and comparing them with the damage functions in the Vrbas river basin, it was possible to conclude the following for a flood-depth of 1 m, as shown in Fig. 5:

- The damage values for the City of Banja Luka (Vrbas river basin) and the region around the River Rhine are similar at around 11% of the assets value.
- The damage values for the municipalities of Gradiška, Jajce, Kotor Varoš, Laktaši and Srbac (Vrbas river basin) are similar to the damage functions used in the study by the US Army Corps of Engineers at around 14% of the assets value.
- The damage values for the municipalities of Bugojno, Čelinac, Donji Vakuf, Gornji Vakuf, Jezero, Šipovo and Mrkonjić Grad (Vrbas river basin) and the region around the River Elba are similar at around 20% of the assets value.

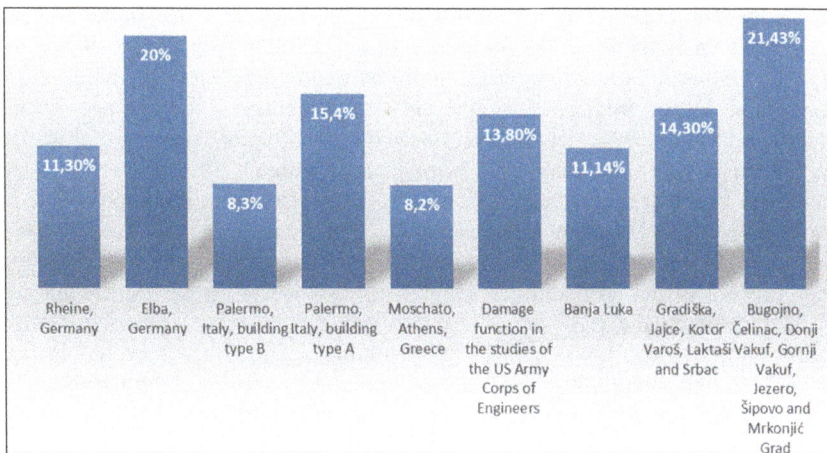

Figure 5:  Comparative values as a damage share of the assets value of households for a flood depth of 1 m. *(Source: Author.)*

## 4 DAMAGE IN THE AGRICULTURAL SECTOR

Damage to agriculture was assessed as the combined damage caused in the flood area for the following agricultural crops: wheat, corn, barley, potato, apple, plum and pear. These are the most represented crops in the target area. They are also the crops for which the statistic institutes in BH collect, analyse and publish data on the planted surface, the yield and similar aspects at the municipal level. The authors of the methodology were aware that there are other crops in the target basin (farming crops, fruits and vegetables), but there are no official records about them per municipalities, so that they are not included in the calculation. The damage to the agricultural crops mentioned above resulted from yield reduction caused by flood, which is presented per hectare in BAM/EUR.

The format of Table 5 was used to calculate the reduction in yield attributed to flooding in each municipality located in the flood area. The reason for presenting data at the municipal level is the significant differences in agricultural development in the various municipalities. For example, Table 5 was used for the City of Banja Luka.

The following table of flood probability was calculated based on the annual maximum flow (m³/s) as per the calculation of the chief technical consultant within the project "Integrating Climate Changes and Reducing the Risk of Flooding in the Vrbas River Basin".

The maximum flood damage (%) compared to the monthly yield, in fact the decrease in yield in percentage compared to the anticipated values, was taken from the second technical report for the agriculture sector in the Vrbas river basin [10]. For the above-mentioned agricultural crops within the category of winter grains (wheat and barley), the maximum flood damage, in fact the decrease in yield in percentage compared to the anticipated value, are attributed to excessive land humidity for a period in excess of 15 days. The same principle applies to the data for periods of excessive humidity for agricultural crops and therefore the period of 15 days was used for corn, 11 days for potato, while 15 days was also used for fruits. Therefore, the results present the maximum damage to agriculture under large water floods. These two values provide weighted loss in percentage calculated as the sum of the flood probability product and the monthly percentage of damage to yields.

The average yield presented as value t/ha is calculated as a 10-year average of yield [11]. Besides, net revenues per crop type are equal to the average price in BAM/kg minus production costs in BAM/kg [12]. The prices of agricultural crops represent the 10-year average [11]. Losses in agricultural crops at the municipal level are presented in BAM/ha and are calculated as a product of weighed loss in percentages, the average yield of the agricultural crops and average net revenue. Based on this equation, the expected flood losses are obtained in BAM per ha of land for certain types of crops.

If there was a cadastre of agricultural crops in BH providing information on the type of crops grown in certain agriculture fields then the values from the next table could have been used to multiply the size of the corps under water by the losses per ha of certain crop type in order to obtain losses per crop and per municipality. However, as we did not know which crops were located in the flood areas, we applied an alternative approach. Geographic information system software was applied to the total flooded surface per municipality. Part of those total flooded surfaces belonged to the above-mentioned agricultural crops. The total land surface under the above-mentioned crops within the entire municipality was then compared to the total size of the municipality. In this way, the share/percentage of land surface for each crop per municipality was obtained and then copied to the flooded areas per municipality. In this way, the sizes of land under crops presented in ha within the flooded area was obtained and calculated.

Table 5:   Calculation of the damage caused to agricultural crops at the municipal level – Banja Luka. (Source: Author.)

| Banja Luka | Months | | | | | | | | | | | | Weighted loss * in % | Average yield in t/ha | Net revenue ** in BAM/kg | Losses in agricultural crops inBAM/ha |
|---|---|---|---|---|---|---|---|---|---|---|---|---|---|---|---|---|
| | 1 | 2 | 3 | 4 | 5 | 6 | 7 | 8 | 9 | 10 | 11 | 12 | (A) | (B) | (C) | (D)=AxBxC |
| Flood probability % | 4 | 12 | 8 | 12 | 18 | 10 | 10 | 2 | 6 | 4 | 4 | 10 | | | | |
| | Maximum flood damage in percentage or the reduction in yield in percentage compared to the anticipated month value | | | | | | | | | | | | | | | |
| Wheat (winter) | 20 | 20 | 50 | 69 | 100 | 100 | 20 | 0 | 0 | 0 | 0 | 0 | 45.5 | 2.7 | 0.074 | 90 |
| Corn | 0 | 0 | 0 | 100 | 100 | 100 | 80 | 70 | 30 | 30 | 0 | 0 | 52.4 | 3.1 | 0.065 | 106 |
| Barley (winter) | 20 | 20 | 50 | 69 | 100 | 100 | 20 | 0 | 0 | 0 | 0 | 0 | 45.5 | 2.5 | 0.080 | 91 |
| Potato | 0 | 10 | 50 | 50 | 100 | 100 | 100 | 100 | 100 | 0 | 0 | 0 | 52.8 | 11.1 | 0.178 | 1,041 |
| Apple | 0 | 0 | 50 | 55 | 73 | 100 | 100 | 50 | 0 | 0 | 0 | 0 | 44.7 | 34.6 | 0.192 | 2,974 |
| Plum | 0 | 0 | 50 | 55 | 73 | 100 | 100 | 50 | 0 | 0 | 0 | 0 | 44.7 | 21.8 | 0.186 | 1,814 |
| Pear | 0 | 0 | 50 | 55 | 73 | 100 | 100 | 50 | 0 | 0 | 0 | 0 | 44.7 | 24.4 | 0.300 | 3,275 |

Note: * Weighted loss in percentages (A) is the sum of the flood probability product and the percentage of damage to crops per month presened in percentages.

** Net revenue in BAM/kg = Average price in BAM/kg – Production costs in BAM/kg.

## 5 CONCLUSION

Flood damage functions have been created for the first time in Bosnia and Herzegovina with a purpose to demonstrate a level of damage caused by floods to household assets and agricultural sector in Vrbas River Basin. These functions have been used to calculate potential flood damages and calculation results have been used (1) as a basis for cost-benefit analysis of flood protection measures; (2) in order to determine what flood risk reduction measures will achieve maximum financial benefits; and (3) to develop insurance products. Obtained values have shown similarities with damage functions developed for other relevant flood damage studies in Europe.

Potential damages have been assessed for residential and agricultural sectors. Hydrological and hydraulic models developed for Vrbas river basin with precise flood depth values for 500-year waters have been used as a base for these calculations. These flood depths and household damage functions were used to determine the average percentage of damage for 1/500 floods. Based on that, the absolute value of assessed damage amounts to approximately BAM 90 million (EUR 46 million). The total average damage value for 1/500 floods for agricultural sector was assessed at approximately BAM 1.4 million (EUR 0.7 million).

The above-mentioned assessments of damage for residential buildings and agriculture sector comply with actual damages. According to the report of the Ministry of Spatial Planning, Construction and Ecology, the floods from 2014., which had the characters of 1/500 floods caused the damage of 130 million BAM (EUR 66 million) [13]. In addition to damages in residential and agriculture sectors, this amount also includes damages in business and public sectors as well.

## REFERENCES

[1]  Smith, D.I., Flood damage estimation – A review of urban stage damage curves and loss functions. *Water SA*, **20**(3), pp. 231–238, 1994.
[2]  Merz, B. et al., Estimation uncertainty of direct monetary flood damage to buildings. *Natural Hazards and Earth System Sciences – Landslide and flood hazards assessment*, **4**, pp. 153–163, 2004.
[3]  Penning-Rowsell, E. et al., *The Benefits of Flood and Coastal Risk Management: A Handbook of Assessment Techniques*, Middlesex University Press: London, 2005.
[4]  Messner, F. et al., *Evaluating Flood Damages: Guidance and Recommendations on Principles and Methods*, HR Wallingford: UK, pp. 26, 2007.
[5]  Davis, S. & Skaggs, L., *Catalogue of Residential Depth-Damage Functions*, U.S. Army Corps of Engineers, Institute for Water Resources, pp. 2, 1992.
[6]  Jovanović, M., Todorović, A. & Rodić, M., Flood risk mapping. *Vodoprivreda*, **41**(1–3), pp. 31–45, 2009.
[7]  Oliveri, E. & Santoro, M., Estimation of urban structural flood damages: The case study of Palermo. *Urban Water*, **2**, pp. 223–234, 2000.
[8]  Pistrika, A., Tsakiris, G. & Nalbantis, I., Flood depth-damage functions for built environment. *Environmental Processes*, **1**, pp. 553–572, 2014.
[9]  U.S. Army Corps of Engineers, *Generic Depth-Damage Relationships for Residential Structures with Basements*, Economic Guidance Memorandum (EGM) 04-01, 2003.
[10] Ćustović, H., Second technical report for agriculture sector in the Vrbas river basin, 2017.
[11] Entity Statistics Institutes, *Annual statistic reports*, 2007–2017.

[12]   *Economics of primary agricultural production and agricultural policy measures in the Federation of B&H*, The Faculty of Agriculture and Food Sciences, University of Sarajevo, Bosnia and Herzegovina, 2010.
[13]   Nezavisne novine. www.nezavisne.com/novosti/bih/Golic-Sliv-Vrbasa-najuredjeniji-u-BiH/565278. Accessed on: 7 Feb. 2020.

# ASSESSMENT OF TRAFFIC DISRUPTION CAUSED BY URBAN FLOODING

DONG HO KANG, KYUNG SU CHOO & BYUNG SIK KIM
Department of Urban and Environmental Disaster Prevention Engineering,
Kangwon National University, South Korea

## ABSTRACT
Recently, local heavy rainfall and typhoons have increased the level of human and property damage in urban environments. Moreover, inundation due to urban flooding has led to severe traffic network disruption and accidents. Rainfall is considered one of the key factors influencing vehicle speed. Thus, in this study, we developed a rainfall (mm)–inundation depth (cm) curve formula to predict changes in vehicle speed under urban flooding. To produce the inundation depth data, rainfall over 10–200 mm was increased by 10 mm, and the respective runoff data were calculated using S-RAT (Spatial Runoff Assessment Tool), a distributed runoff model. Next, the simulated runoff data was inputted to the flood inundation model, Flo-2D. With the use of such assessed formula and previous research on the rainfall (mm)–vehicle speed (km/h) relation, the rainfall (mm)–inundation depth (cm)–vehicle speed (km/h) curve formula was created. The pilot areas were Seoul metropolitan in South Korea, which was divided into 1×1 km grids. Finally we could analyse the impact of urban flooding on traffic disruption. According to the analysis, the flooding area during the time of the actual rainfall was found to have poor traffic, but there are some areas where the speed was not reduced. The results from the flood depth–vehicle speed curve could be derived by reflecting the hydrological characteristics of each grid, but the curves for the rainfall–vehicle speed formula did not reflect the characteristics, resulting in most grids having similar patterns. Through this process, a traffic reduction rate map was developed and the speed reduction by section was suggested.
*Keywords: urban flood, rainfall (mm)–flood depth (m) curve, rainfall (mm)–flood depth (m)–vehicle speed (km/h) curve, traffic disruption, impact assessment.*

## 1 INTRODUCTION
Recent advances in technology have made it possible to provide rapid and accurate information to the public by obtaining more reliable and widely spread rainfall data and processing and analyzing collected rainfall data through various equipment. Furthermore, studies are being actively conducted to analyze places where a lot of vehicles are related and to recommend destinations through different routes through real-time data. However, studies on the traffic situation prediction using rainfall data and studies on the speed of roads in flooded urban roads are also lacking. The purpose of this study is to analyze the relationship between flood depth–vehicle speed rather than rainfall–vehicle speed. Later, a risk map can be created to help the route bypass if a rain event occurs and the road is flooded. In this study, the rainfall–flood depth–vehicle speed curve was calculated using the rainfall–flood depth curve and the flood depth–vehicle speed curve. In 2011, the rainfall was inspected for flooding in the Sadang-dong area of Seoul.

Jeong et al. [1] analyzed the traffic characteristics according to precipitation using RWIS (Road Weather Information System) data and detector data, and suggested that speed and traffic volume decreased when rainfall occurred. Speed estimation curves according to rainfall intensity were presented, and the criteria for low and medium precipitation were 0.4 mm/5 min, and the criteria for moderate and strong precipitation were 0.8 mm/5 min when rain levels were classified using the reduction in speed according to rainfall. Lam et al. [2] proposed a speed–density relation under various rainfall conditions considering vehicle speed, speed according to traffic volume, and traffic volume that can be accommodated by

WIT Transactions on The Built Environment, Vol 194, © 2020 WIT Press
www.witpress.com, ISSN 1743-3509 (on-line)
doi:10.2495/FRIAR200121

road, and suggested a function formula of vehicle speed and traffic volume according to rainfall intensity. In Hong Kong and other Asian cities, it is suggested that rainfall intensity should be considered in the design, operation, and evaluation of road facilities because the average annual rainfall intensity is relatively high. Mashros et al. [3] estimated the effect of rainfall on the moving speed and the extent of the speed decrease using three months of data. Thus, the rainfall intensity is divided into four stages and a graph shows that the moving speed decreases in percentiles of 15, 50, and 85 as the rainfall intensity increases. Kim and Oh [4] used the traffic volume data of the general national road and the rain data of the Korea Meteorological Administration to conduct a study on the daily mean traffic variation by rainfall intensity (weekly, weekend) and suggested that the more the rainfall, the more the average daily traffic varies by up to 5.48% in all days.

## 2 THEORETICAL BACKGROUND

### 2.1 Rainfall–flood depth curve

The rainfall–flood depth curve is a curve using the flood depth, which is the result of the rainfall data provided by the Korea Meteorological Agency and the modeling results. In order to calculate the depth of flooding, first a rainfall runoff model was used with the rainfall of 10–200 mm as the Huff quartile method. The estimated runoff is applied to the flood overflow model and the rainfall–flood depth curve is created using the flood depth data. See Lee et al. [5] for further analysis of the rainfall–depth curve.

### 2.2 Rainfall–flood depth–vehicle speed curve

In this study, the flood depth–vehicle speed curve was obtained by using the data from a previous analysis between the flood depth and the speed of the vehicle. The estimation function data presented in Pregnolato et al. [6] were used to calculate the flood depth–vehicle speed curve. This lead to the linear regression equation

$$y = 85 \times e^{-9x}(R^2 = 0.87). \tag{1}$$

## 3 APPLICATION TO THE STUDY BASIN

### 3.1 Study basin

To carry out this study, Seoul City, the most urbanized city in Korea, was selected as the target area. The city area was divided into 1 km × 1 km grids along the shrine, Bangbae, and Seocho-dong, which were affected by a heavy rain of about 300 mm from 1:00 am to 11:00 pm on 27 July 2011. Fig. 1 shows the area of Sadang-dong, which is divided into grids. Fig. 2 shows the grid number, road network, and speed limit for each grid in Sadang-dong. In this study, we analyzed grids 23 and 26 where actual damage had occurred in the past.

### 3.2 Rainfall–flood depth curve calculation results

In order to calculate the rainfall–flood depth curve, the area of Sadang-dong was divided into 1 km × 1 km grid and numbers were assigned to each grid square. The vehicle speed was set based on the road speed limit of each grid square. Most of them included roads in downtown Seoul, indicating that there were many 60 km speed limit roads. Figure 3 shows the calculation of the rainfall-depth curve in grid 23 and 26 where the damage occurred.

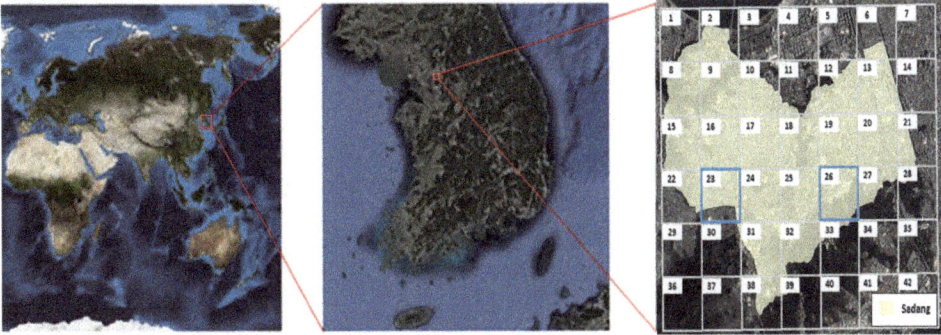

Figure 1:  Study basin.

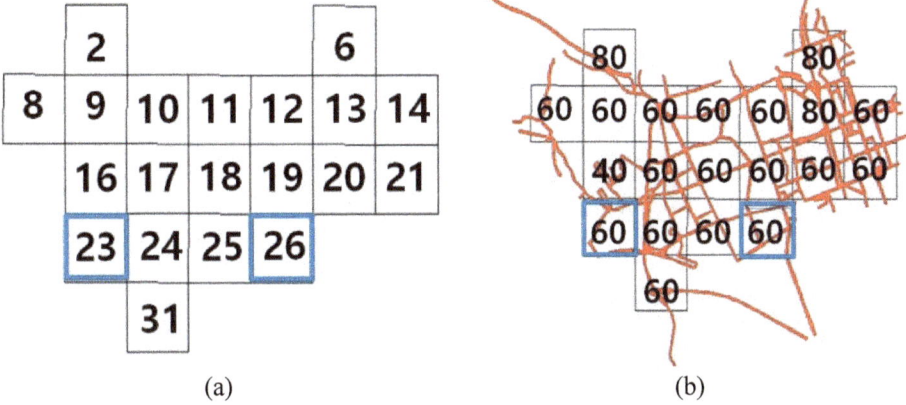

Figure 2:    Sadang-dong grid number and speed limit for each grid. (a) Grid number; and (b) Speed limit for each grid.

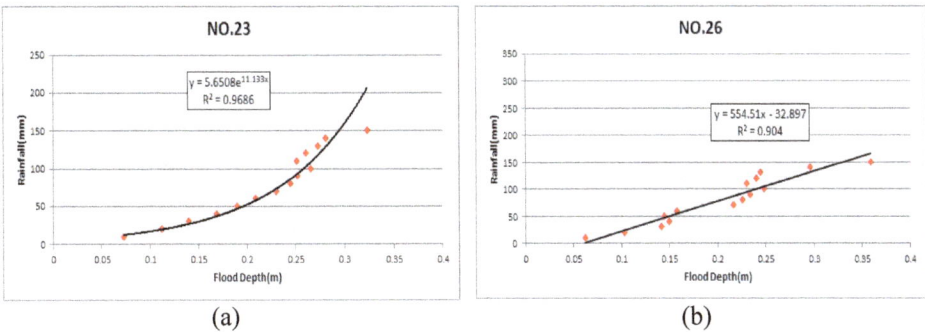

Figure 3:  Result of rainfall-depth curve.

### 3.3 Results of estimation of rainfall–flood depth–vehicle speed curve

On 27 July 2011, there was a traffic restriction due to flooding at 6:30 am at the Sadang Station included in the 23rd grid, and traffic resumed at 10:00 am. As a result of the rainfall–flood depth–vehicle velocity curve calculation, it was predicted that a vehicle could not travel at the time when the actual damage occurred, and it can be confirmed that the traffic resumes after damage time. Figure 4 shows the rainfall–flood depth–vehicle curve result.

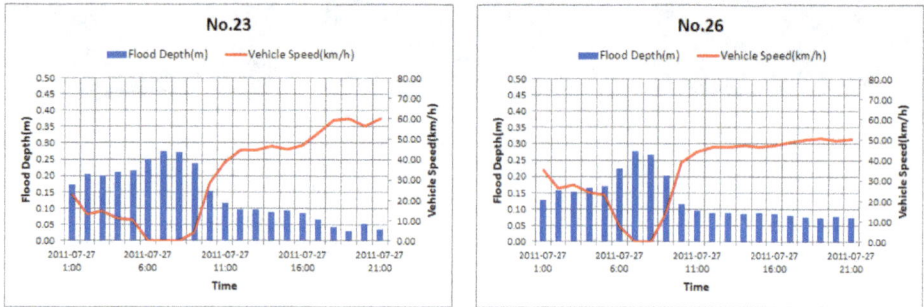

Figure 4:  Rainfall–flood depth–vehicle speed curve.

### 3.4 Traffic disruption impact assessment by grid

The rate of reduction of vehicle speed by grid was calculated using the calculated rainfall–flood depth–vehicle speed curve. Traffic disruption impact assessment was carried out by dividing the rate of reduction of vehicle speed into four levels of very low, low, medium and high. The criteria for dividing into four grades were calculated as the rate of decrease according to the speed limit for each grid. Very low has a reduction rate of 0–25%, low 25–50%, medium 50–75%, and high 75–100%. Fig. 5 is a picture of the traffic disruption rating of the time when the vehicle was controlled by waterlogging at 7:00 am on 27 July 2011 and the traffic disruption rating at 1:00 pm when vehicle control was lifted. It can be seen that the

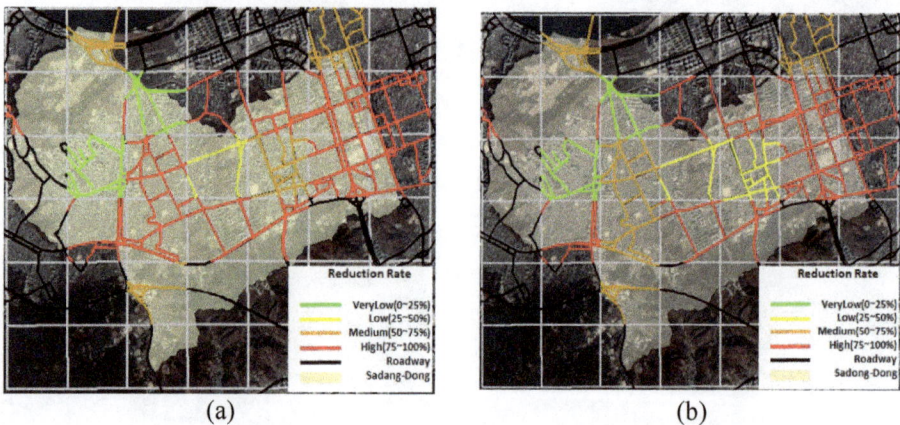

Figure 5:  Analysis of impact assessment. (a) 7:00 am on 27 July 2011 traffic disruption rating; and (b) 9:00 am on 27 July 2011 traffic disruption rating.

vehicle reduction rate is high in many grids at the time of the volume control, and that there is a grid square having smooth traffic due to the low vehicle reduction rate. This can allow bypassing roads with high rates of reduction so that drivers can reach their destination.

## 4 CONCLUSION

In this study, the rainfall–flood depth–vehicle speed curve was calculated using the rainfall–flood depth curve and the flood depth–vehicle speed curve. This curve is generated by time unit and the result can be confirmed. For the verification of the calculated curves, the test was conducted using the 27 July 2011 event, at which actual damage occurred in the past.

The results of the verification confirmed that traffic was not possible in the rainfall–flood depth–vehicle speed curve from 6:00 am when actual vehicle traffic was controlled until 10:00 am when traffic was resumed. In this study, only two grid squares were used for model verification, but if there are additional articles, news, and damage pictures that actually occurred, it is necessary to further verify the model. Rainfall–flood depth–vehicle speed curves still need a lot of complementary work. It is expected to increase the practical utilization if the individual's psychology and behavioral patterns such as the driver's understanding the characteristics of the rain and thus further develop the relational expression.

Using a proven curve, the traffic disturbance class by time zone was determined to be four stages from 1:00 am to 11:00 pm on 27 July 2011, and there was a smooth traffic grid square during the time when vehicle control was carried out due to flooding as a result of the calculation, which can be utilized in various ways in the future.

If a public servant uses this study, it is expected that the budget for road maintenance can be organized according to the proportion of road connectivity (where most pass by when vehicles go elsewhere) by upgrading each road characteristics and drain pipe network.

In this study, the depth of flooding can be calculated and the vehicle speed can be analyzed according to the predicted rainfall. It is still in the early stages of development, but, it is believed that can continue if the recommended route can be taken into account by considering the speed of the vehicle and the minimum cost, and if actual flooding can be used to identify its effect.

## ACKNOWLEDGEMENTS

This work was funded by the Korea Meteorological Administration Research and Development Program under grant KMI[2018-03010]. This work was supported by Korea Environment Industry and Technology Institute (KEITI) through Advanced Water Management Research Program, funded by Korea Ministry of Environment (MOE) (83091). This paper was financially supported by Ministry of the Interior and Safety as "Human Resource development Project in Disaster Management".

## REFERENCES
[1] Jeong, E., Oh, C. & Hong, S., Prediction of speed by rain intensity using road weather information system and vehicle detection system data. *Intelligent Transport Systems*, **12**(4), pp. 44–55, 2013.
[2] Lam, W.H.K., Tam, M.L., Cao, X. & Li, X., Modeling the effects of rainfall intensity on traffic speed, flow, and density relationships for urban roads. *Journal of Transportation Engineering*, **139**(7), pp. 758–770, 2013.
[3] Mashros, N., Ben-Edigbe, J., Alhassan, H.M. & Hassan, S.A., Investigating the impact of rainfall on travel speed. *Journal Teknologi*, **71**(3), pp. 33–38, 2014.

[4] Kim, T.W. & Oh, J.S., Analysis of provincial road in national highway average speed variation according to rainfall intensity. *Journal of the Korea Contents Association,* **15**(4), pp. 510–518, 2015.

[5] Lee, S.H., Kang, D.H. & Kim, B.S., A study on the method of calculating the threshold rainfall for rainfall impact forecasting. *Journal of the Korean Society of Hazard Mitigation,* **18**(7), pp. 93–102, 2018.

[6] Pregnolato, M., Ford, A., Wilkinson, S.M. & Dawson, R., The impact of flooding on road transport: A depth-disruption function. *Transportation Research Part D: Transport and Environment,* **55**, pp. 67–81, 2017.

# SECTION 5
# BLUE-GREEN
# INFRASTRUCTURE

# INFILTRATION–EXFILTRATION SYSTEMS DESIGN UNDER HYDROLOGICAL UNCERTAINTY

ANITA RAIMONDI, MARIANA MARCHIONI, UMBERTO SANFILIPPO & GIANFRANCO BECCIU
Politecnico di Milano, Italy

## ABSTRACT

The current scenario of urban drainage management encourages the use of sustainable urban drainage systems (SUDSs) acting on runoff volume, peak flow and quality standards. Infiltration–exfiltration systems (IES) are part of SUDSs: they are composed of a high permeability surface and a structure that functions as a reservoir; depending on site characteristics, stormwater can be infiltrated on subsoil, drained or a combination of both. IES design consists basically of sizing the reservoir layer depth according to rainfall, soil and drainage characteristics considering a maximum drainage time between 24 and 72 hours. If long records of rainfall data are available, continuous simulations are performed, otherwise the so called "design event" method is used. It considers a single rainfall event, neglecting the possibility that the storage capacity can be partially filled from previous rainfalls. This paper proposes an analytical probabilistic approach for IES design, combining the simplicity of "design event" methods and the statistical reliability of continuous simulations. It considers the possibility of pre-filling of the reservoir layer from more than one previous event, a key aspect for the correct design of low release structures as IES. The approach has been tested applying proposed equations to a case study, in Milan, Italy, and comparing the results with those from the continuous simulation of recorded data.

*Keywords: infiltration–exfiltration systems, hydrological uncertainty, sustainable urban drainage systems, pre-filling, analytical probabilistic modelling.*

## 1 INTRODUCTION

A characteristic of climate change is the increases of heavy rainfalls characterized by short durations and high intensities. These phenomena, especially on urbanized catchments, causes frequent flood and considerable economic damages. The increase of impervious surfaces, typical of urban areas, decreases soil infiltration and evapotranspiration, so increasing the runoff volume drained to the stormwater drainage system. To avoid the overload of the networks in last decades stormwater source management is encouraged and sometimes mandatory. SUDSs allow retention or detention of stormwater on its source, reducing peak flows, runoff volumes and in some cases the pollution load also providing amenity and biodiversity opportunities into the urban context. Despite their benefits, SUDSs are not always easy to implement and develop in city centre characterized by a high urban density, where retrofitting is generally more expensive and may be limited to few urban spaces. The use of permeable pavements can be effective because it doesn't require any additional space, but it is generally limited to car parks and low traffic roads (Marchioni and Becciu [1]).

IESs are a good trade-off to limit the adaptations to road gutters, that are less stressed by dynamic loads; they are linear street side channels composed of a permeable surface layer and a gravel and sand layer and an underdrain if necessary. Their use may achieve several goals: limiting runoff discharged into the drainage system, since part of rainwater can be infiltrated; reducing peak flow into the network for the temporary storage inside the permeable layers; removing pollutants through filtration, sedimentation, adsorption, biodegradation and volatilization. Moreover, IES avoid ponding, promote aquifer recharge and reduce inlet maintenance.

WIT Transactions on The Built Environment, Vol 194, © 2020 WIT Press
www.witpress.com, ISSN 1743-3509 (on-line)
doi:10.2495/FRIAR200131

Initial researches on permeable pavements in 1970s were mostly conducted on laboratory normally using rainfall simulation (Pratt [2]), while first full-scale tests were performed starting from the 1980s (Hogland et al. [3], Pratt [4], Pratt et al. [5], Legret et al. [6], Legret and Colandini [7], Pagotto et al. [8], Asaeda and Ca [9], Schlüter and Jefferies [10], Brattebo and Booth [11], Dreelin et al. [12], Morgenroth et al. [13], Newman et al. [14]). Results from these studies show a great variability in stormwater capture efficiency, strictly due to differences in design and climate conditions.

With reference to IES, different research analysed their performances, testing their potential effectiveness in terms of peak flow and volume reduction (Marchioni and Becciu [15]) and pollutant load removal (Teng and Sansalone [16], Sansalone and Teng [17]). Traditional methods, based on design storm event, are unreliable because they just neglect the possibility that the structure is partially pre-filled and that, therefore, the whole capacity is not available at the beginning of the design event.

In this paper, an analytical probabilistic model to analyse the efficiency of IES in terms of their capability to cope with ponding along streets sides has been developed. This kind of models have been proposed at first by Eagleson [18], [19] and Adams and Papa [20] as alternative to continuous simulations to model rainfall-runoff transformation. They have been applied to urban drainage systems to analyse runoff volume and floods peaks (Guo and Adams [21], [22], Guo et al. [23]), stormwater detention storages (Guo and Adams [24], [25], Bacchi et al. [26], Balistrocchi et al. [27], Raimondi and Becciu, [28], [29]; Becciu and Raimondi [30]–[33]) and recently have been used to estimate the efficiency of SUDSs; in particular, analytical probabilistic approaches have been used to analyse green roofs (Zhang and Guo [34], [35], Guo [36], Raimondi and Becciu [37]), rainwater harvesting systems (Guo and Baetz [38], Raimondi and Becciu [39]–[41], Becciu et al. [42]), infiltration trenches (Guo and Gao [43]), bioretention systems (Zhang and Guo [44]), permeable pavements (Zhang and Guo [45]).

These models derive analytical equations of the variable of interest from the probabilistic distribution function (PDF) of rainfall event characteristics and the mathematical representations of the hydrologic processes; they have the great advantage that it is easy to apply them to different kinds of structures under different climate conditions.

Their limitation is that just two rainfall events have been considered and the system has been assumed fully filled at the end of the first event of the current cycle (Howard [46], Loganathan and Delleur [47], Adams and Papa [20]).

Raimondi and Becciu [28] discussed the number of chained events to be considered in the model: they concluded that, for long IETD and high outflow rates, two chained rainfall events may be acceptable; for low outflows facilities or when strict limitations on discharges in the downstream drainage system are imposed, Authors suggested to assume three chained rainfall events to consider the contribution to outflow of pre-filling volumes from previous events.

In Raimondi and Becciu [37], the authors developed a model to consider a chain of N rainfall events; it has been successfully applied to green roofs and in this paper has been tested on IESs.

The model adapted to an IES is then applied to a case study; the influence of simplifying assumptions of the model on results has been tested and discussed; the accuracy of the proposed approach has been validated by comparing results from analytical equations with those obtained from the continuous simulation of real data.

## 2 HYDROLOGICAL MODELING

IESs are placed on road gutters to receive runoff from streets and infiltrate it into the underlying layer, to the native soil and/or the drainage system. A typical IES is composed of a permeable surface layer, followed by a gravel aggregate layers that function as a reservoir. A general scheme of the system has been proposed on Fig. 1.

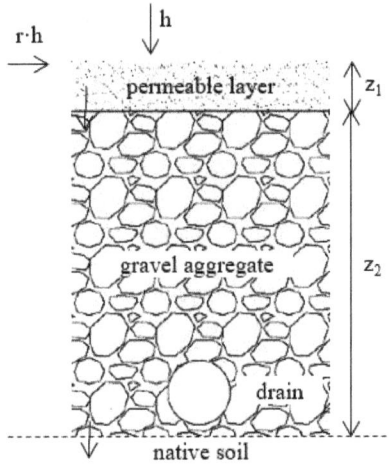

Figure 1:  Scheme of an infiltration–exfiltration system.

The permeable surface layer must allow water to infiltrate; therefore, it has a high porosity structure with open and interconnected pores where water and air can pass through; infiltration must be fast enough to avoid the possibility of significant ponding for most of the rainfall events. Both porous asphalt and pervious concrete are suitable for this layer. The gravel aggregate layer must have a high void rate, to perform as a reservoir. But the high void content results on less strength; for this reason, IES are normally applied in street sides. The gravel aggregate layer is usually equipped with an overflow control device so that the water level inside the stone reservoir cannot rise to the pavement level or the surface of the IES during any large storm events. For systems without underdrains, the in-situ soil needs to be highly permeable and with low clay contents (generally less than 30%, U.S. EPA [48]).

Surface runoff rarely occurs, thanks to the high permeability of the surface layer (Brattebo and Booth [11], Collins et al. [49]). Inflow to IES can be trapped by small depressions on the surface or adsorbed by the permeable surface layer; the rest is infiltrated into the gravel aggregate layer, and from its bottom can percolate to the underlying soil. When inflow rate into gravel aggregate layer exceeds infiltration capacity of natural soil, storage occurs, and water level of stone reservoir rises. If storage capacity is then fulfilled, rainwater can either be drained away through underdrain (if installed in the system) or flow away over the surface of the system as surface outflow. After a rainfall event has come to the end, rainwater stored into the IES is depleted by both percolation through the bottom of the gravel aggregate layer and evaporation.

Water balance for an IES is represented by the following equation, usually expressed in millimetres of water over the system's surface area:

$$I = F + W + R,$$    (1)

where $I$ represents the inflow into the structure, $R$ the outflow from the IES, $F$ the infiltration into the underlaying soil and $W$ the stored rainwater volume. Evaporation is neglected in eqn (1) since it is generally negligible in comparison to infiltration (Nemirovsky et al. [50]). If the system is equipped with underdrains, the outflow $R$ is the sum of the volumes of the surface outflow and drain outflow; for IESs without underdrains $R$ is the volume of surface outflow from the system.

Rainwater stored into the IES can vary between zero and $W_{max}$, that is the retention capacity of the system. For IESs without underdrains, retention capacity $W_{max}$ consists of three parts: surface depressions and void space of the permeable surface layer, and void space of the gravel aggregate layer.

In an IES with underdrain, the stormwater held in the void space of the surface permeable layer and in the void space of gravel aggregate layer which is above the underdrains can be quickly drained away through the underdrains. Therefore, the retention capacity of the system only consists of the surface depressions plus the void space of the part of the stone reservoir which is below the underdrains.

Inflow to IES $I$ incudes surface runoff from contributing impervious areas $(r \cdot h)$ and rainwater directly falling onto the structure $(h)$:

$$I = h + r \cdot h, \tag{2}$$

where $r$ = ratio between the contributing impervious area and the permeable IES area.

Here, for simplicity, it has been considered that the whole rainfall from the contributing area is collected into the IES. Moreover, despite its random nature (Becciu and Paoletti [51]), the runoff coefficient has been assumed equal to one (completely impervious contributing area).

In addition, infiltration capacity of surface permeable layer and gravel aggregate layer has been assumed to be always greater than inflow rate into the system.

To test the effectiveness of IES as SUDS, an analytical probabilistic model has been proposed. Basically, it consists in estimating the Probability Distribution Functions (PDFs) of the variable of interest from the PDFs of the input variables, which are the rainfall event characteristics, coupled to the mathematical representations of the hydrologic processes occurring in the IES.

Input rainfall variables are rainfall depth, rainfall duration and interevent time; they are considered independent and exponentially distributed. To identify independent events from a continuous series of rainfalls, a minimum interevent time has been defined (IETD). If interevent time between two consecutive rainfall events is lower than IETD then the two storms are joined in a single event, otherwise they are considered independent. In literature, different methods to select IETD have been proposed: estimating the autocorrelation coefficient of observations sample, choosing the values for which the coefficient of variation tends to one, evaluating the relationship between IETD and the average number of rainfall events. In practice, IETD must be related to catchment response characteristic; generally, shorter IETD are suggested for small urban catchment with quick concentration times while for large rural catchment IETD can be also of several hours.

Different studies highlighted as exponential PDF provides a good fit to frequency histograms of main rainfall characteristics (Adams et al. [52], Eagleson [18], Bedient and Huber [53]). Bacchi et al. [26] tested that, for most of the Italian basins, the Weibull probability distribution function better fits the frequency distribution of meteorological input variables than the exponential probability distribution function; however, its use would involve a considerable complication in the integration of the equations. Becciu and Raimondi [30] verified that the double-exponential probability distribution function well fits the frequency distribution of observed data for main rainfall characteristic parameters; such distribution may be easily integrated but derived expressions are more complex. Moreover, its application to a case study highlighted that the use of the double-exponential probability distribution function little improves the accuracy of results and that the bias due to the use of the exponential probability distribution function is. In particular, the PDFs of rainfall depth, rainfall duration and interevent time are, respectively:

$$f_h = \xi \cdot e^{-\xi \cdot h}, \tag{3}$$

$$f_\theta = \lambda \cdot e^{-\lambda \cdot \theta}, \tag{4}$$

$$f_d = \psi \cdot e^{-\psi \cdot (d - IETD)}, \tag{5}$$

where $\xi = 1/\mu_h$; $\lambda = 1/\mu_\theta$; $\psi = 1/(\mu_d - IETD)$. The parameters $\mu_h$, $\mu_\theta$ and $\mu_d$ are the mean values of respectively rainfall depth, rainfall duration and interevent time.

## 3  METHODOLOGY

For a correct design of an IES, outflow discharge should be avoided (full underlying soil infiltration) or limited keeping into account the downstream network capacity and/or the discharge limitations imposed by law and regulations. To estimate overflow probability from an IES, water content in the system at the end of a generic rainfall event must be calculated. The computational scheme can be summarized as follows:

$$W_i = \begin{cases} W_{i-1} - f \cdot d_i + I_i - f \cdot \theta_i & Condition_1 \\ I_i - f \cdot \theta_i & Condition_2 \\ W_{max} & Condition_3; Condition_4 \\ 0 & Otherwise \end{cases} \tag{6}$$

$Condition_1: W_{i-1} - f \cdot d_i > 0; \; 0 < W_{i-1} - f \cdot d_i + I_i - f \cdot \theta_i < W_{max}$
$Condition_2: W_{i-1} - f \cdot d_i \leq 0; \; 0 < I_i - f \cdot \theta_i < W_{max}$
$Condition_3: W_{i-1} - f \cdot d_i \leq 0; \; I_i - f \cdot \theta_i \geq W_{max}$
$Condition_4: W_{i-1} - f \cdot d_i > 0; \; W_{i-1} - f \cdot d_i + I_i - f \cdot \theta_i \geq W_{max},$

for $i = 1, \dots, N$ where $N$ is the number of considered rainfall events.

Water content for $i = 0$, that is $W_0$, is:

$$W_0 = \begin{cases} I_0 - f \cdot \theta_0 & 0 < I_0 - f \cdot \theta_0 < W_{max} \\ W_{max} & I_0 - f \cdot \theta_0 \geq W_{max} \\ 0 & I_0 - f \cdot \theta_0 \leq 0 \end{cases}. \tag{7}$$

With reference to eqn (7):

*Condition₁* expresses the case of pre-filling from previous rainfall event at the end of the considered event and this does not produce runoff.

*Condition₂* expresses the case in which there is no pre-filling from previous rainfall event at the end of the considered event and this does not produce runoff.

*Condition₃* expresses the case in which there is no pre-filling from previous rainfall event at the end of the considered event and this produces runoff.

*Condition₄* expresses the case of pre-filling from previous rainfall event at the end of the considered event and this produces runoff.

Variable $f$ in eqns (7) and (8) represents infiltration rate; it has been assumed constant and equal to the saturated hydraulic conductivity of the native soils.

Overflow from an IES is a random variable strictly depending on rainfall characteristics, infiltration rate of native soils, and maximum storage capacity of the system. It results:

$$R_i = \begin{cases} W_{i-1} - f \cdot d_i + I_i - f \cdot \theta_i - W_{max} & \textit{Condition}_1 \\ I_i - f \cdot \theta_i - W_{max} & \textit{Condition}_2; \textit{Condition}_3 \\ W_{max} - f \cdot d_i + I_i - f \cdot \theta_i - W_{max} & \textit{Condition}_4 \\ 0 & \textit{Otherwise} \end{cases} \quad (8)$$

$\textit{Condition}_1\colon W_{i-1} < W_{max};\ W_{i-1} > f \cdot d_i;\ W_{i-1} - f \cdot d_i + I_i - f \cdot \theta_i > W_{max}$

$\textit{Condition}_2\colon W_{i-1} < W_{max};\ W_{i-1} \le f \cdot d_i;\ I_i - f \cdot \theta_i > W_{max}$

$\textit{Condition}_3\colon W_{i-1} \ge W_{max};\ W_{max} \le f \cdot d_i;\ I_i - f \cdot \theta_i > W_{max}$

$\textit{Condition}_4\colon W_{i-1} \ge W_{max};\ W_{max} > f \cdot d_i;\ W_{max} - f \cdot d_i + I_i - f \cdot \theta_i > W_{max},$

for $i = 1, \dots, N$ where $N$ is the number of considered rainfall events.

Outflow for $i = 0$, that is $R_0$, results:

$$R_0 = \begin{cases} I_0 - f \cdot \theta_0 - W_{max} & I_i - f \cdot \theta_i > W_{max} \\ 0 & \textit{Otherwise} \end{cases}. \quad (9)$$

With reference to eqn (9):

*Condition₁* expresses the case in which there is no outflow at event $i - 1$, there is pre-filling from event $i - 1$ at the beginning of event $i$ and there is runoff from the IES at the end of event $i$;

*Condition₂* expresses the case in which there is no outflow from event $i - 1$, there is no pre-filling from event $i - 1$ at the beginning of event $i$ and there is runoff from the IES at the end of event $i$;

*Condition₃* expresses the case in which there is outflow from event $i - 1$, there is no pre-filling from event $i - 1$ at the beginning of event $i$ and there is runoff from the IES at the end of event $i$;

*Condition₄* expresses the case in which there is outflow from event $i - 1$, there is pre-filling from event $i - 1$ at the beginning of event $i$ and there is runoff from the IES at the end of event $i$.

With reference to the derived probability distribution theory (Benjamin and Cornell [54]), outflow PDF from the system can be derived from PDFs of rainfall depth $h$, rainfall duration $\theta$ and interevent time $d$. It has been estimated setting $h = h_i = h_{i+1}$, $\theta = \theta_i = \theta_{i+1}$, $d = d_i = d_{i+1}$ in eqn (9); this leads to exclude *Condition₂*.

Outflow probability have been estimated distinguishing two different conditions: maximum emptying time, that is time needed to empty the retention capacity when it is full, respectively lower and higher than the minimum interevent time *IETD*:

- for Case 1, the pre-filling from previous rainfalls at the beginning of the considered event has been excluded with the full storage capacity available;
- for Case 2, the possibility that retention volume is partially filled from previous rainfalls has been analyzed.

The term $IA$ in eqns (2) and (3), representing the stormwater trapped by small surface depressions, has been neglected, since it is generally very low. a threshold outflow volume $\bar{R}$, related for example to downstream discharge constrains into the drainage system, has been used in the calculation.

Case 1: $W_{max}/f \le IETD$:

$$P_R = P(R > \bar{R}) = \int_{h = \frac{W_{max} + \bar{R} + f \cdot \theta}{1 + r}}^{\infty} p_h \cdot dh \int_{\theta = 0}^{\infty} p_\theta \cdot d\theta = \gamma \cdot e^{-\frac{\xi}{1+r}(W_{max} + \bar{R})}, \quad (10)$$

with: $\gamma = \frac{\lambda \cdot (1+r)}{f \cdot \xi + \lambda \cdot (1+r)}$.

Case 2: $W_{max}/f > IETD$:

$$P_R = P(R > \bar{R}) = \int_{\theta=0}^{\infty} p_\theta \cdot d\theta \int_{h=\frac{W_{max}+\bar{R}+f\cdot\theta}{1+r}}^{\infty} p_h \cdot dh$$

$$+ \sum_{i=2}^{N} \left[ \int_{\theta=0}^{\infty} p_\theta \cdot d\theta \int_{d=IETD}^{\frac{W_{max}+\bar{R}\cdot(1-i)}{f}} p_d \cdot dd \int_{h=\frac{W_{max}+\bar{R}+(i-1)\cdot f\cdot d}{i\cdot(1+r)}+\frac{f\cdot\theta}{1+r}}^{\frac{W_{max}+(i-2)\cdot f\cdot d}{(i-1)\cdot(1+r)}+\frac{f\cdot\theta}{1+r}} p_h \cdot dh \right]$$

$$= \gamma \cdot \left\{ e^{-\frac{\xi}{1+r}(W_{max}+\bar{R})} - \psi \cdot (1+r) \cdot \sum_{i=2}^{N} \left[ (i-1)\cdot\beta_i \cdot e^{-\frac{\xi}{(1+r)\cdot(i-1)}[f\cdot IETD\cdot(i-2)+w_{max}]} + \right. \right.$$

$$i \cdot \beta_i^* \cdot e^{-\frac{\xi}{i\cdot(1+r)}[\bar{R}+W_{max}+f\cdot IETD(i-1)]} + \xi \cdot f \cdot \beta_i \cdot \beta_i^* \cdot$$

$$\left. \left. e^{\psi\cdot IETD+\frac{\psi}{f}[\bar{R}\cdot(i-1)-w_{max}]-\frac{\xi}{1+r}(2\cdot\bar{R}+w_{max}-i\cdot\bar{R})} \right] \right\},$$

(11)

where: $\beta_i = \frac{1}{\xi\cdot f\cdot(i-2)+\psi\cdot(i-1)\cdot(1+r)}$; $\beta_i^* = -\frac{1}{i\cdot\psi\cdot(1+r)+(i-1)\cdot\xi\cdot f}$.

## 4 CASE STUDY

The proposed IES model has been applied to a case study considering the cross section indicated on Fig. 1 and Milan (Italy) rainfall data. The system has a surface permeable layer of depth $z_1 = 10\ [cm]$ and porosity $n_1 = 0.15\ [-]$; a gravel aggregate layer of depth $z_2 = 60\ [cm]$ and porosity $n_2 = 0.35\ [-]$ has been considered. So, the maximum retention capacity results equal to $W_{max} = z_1 \cdot n_1 + z_2 \cdot n_2 = 21.5\ [cm]$. Infiltration capacity of the underlying natural soil has been estimated equal to $f = 3.6\ [mm/hour]$ (it corresponds to a sandy-clay loam with hydraulic conductivity equal to $K = 10^{-6}\ [m/s]$). The ration between the impervious contributing area and the infiltration area has been assumed equal to $r = 4$. Input rainfalls are those recorded at Milano-Monviso gauge station in the period 1971–2005. To identify independent rainfall events from the continuous data records, a $IETD = 1\ [hour]$ has been used. The main characteristics of rainfall variables mean $\mu$, standard deviation $\sigma$ and coefficient of variation $V$, are reported in Table 1.

Table 1:   Main characteristics of rainfall variables.

|  | μ | σ | V = σ/μ |
|---|---|---|---|
| h (mm) | 7.62 | 12.40 | 1.63 |
| θ (hour) | 4.32 | 5.79 | 1.34 |
| d (hour) | 66.50 | 129.00 | 1.94 |

The hypothesis of exponential PDF is not perfectly suitable to the experimental data, especially for rainfall depth and interevent time, but the effects on the results have been deeply tested by the authors in [28], concluding that the bias due to its use can be considered negligible. Table 2 contains the correlation coefficients among the three hydrological parameters. Interevent time results are just weakly correlated to the other two variables, while the correlation between rainfall depth and duration is quite high.

Table 2: Correlation index among rainfall variables.

| | |
|---|---|
| $\rho_{h,d}$ (-) | 0.01 |
| $\rho_{\theta,h}$ (-) | 0.70 |
| $\rho_{d,\theta}$ (-) | 0.01 |

Eqn (11) has be tested considering the maximum retention capacity $W_{max}$ varying between zero to 250 $[mm]$ (Fig. 2); the threshold retention capacity $\bar{R}$ has been set equal to zero. Outflow PDF has been estimated considering a single event $i = 1$, a couple of events $i = 2$ and four chained events $i = 4$; the results have been compared with observed frequency calculated considering the whole series of records. The approach of considering a chain of events instead of one improves fitting, since pre-filling from previous rainfall has been considered in the model. But the proposed equation overestimates the probability of overflow for low retention capacities because the effects of the simplifying assumptions are more remarkable. A retention capacity equal to $W_{max} = 215 \ [mm]$, that is the case of the considered IES, corresponds to a return period $T = 34 \ [years]$ (that can be considered acceptable for design purposes). If a return period of $T = 50 \ [years]$ is considered, the overflow probability results equal to 0,02 $[-]$, that corresponds to a retention capacity equal to $W_{max} = 270 \ [mm]$. Always considering a surface permeable layer of depth $z_1 = 10 \ [cm]$ and porosity $n_1 = 0.10$, the gravel aggregate layer should be equal to $z_2 = 73 \ [cm]$, setting $n_2 = 0.35$.

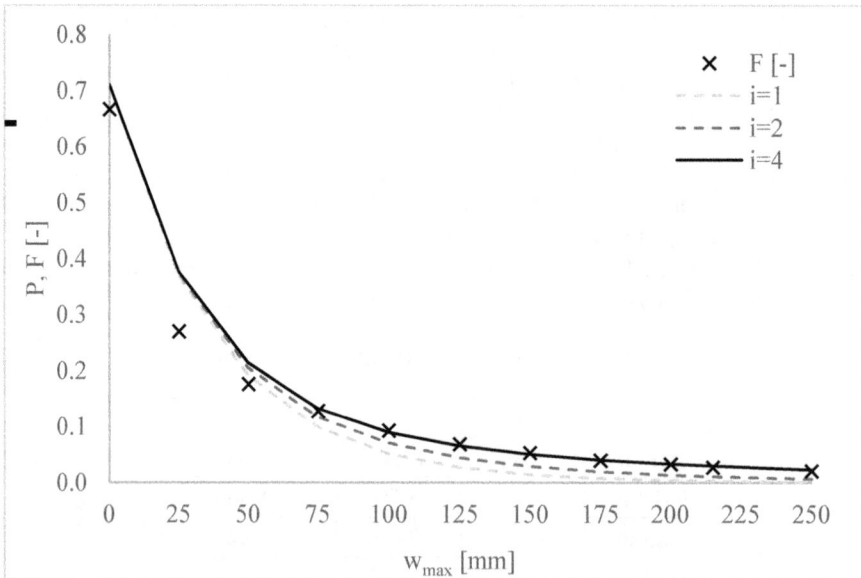

Figure 2: Outflow PDF from an IES varying retention capacity.

## 5 CONCLUSIONS

IESs can be an effective tool for SUDSs, especially in cases where the presence of vehicular traffic prevents the use of permeable pavements for the whole considered area. The proposed approach shows good results, compared with frequency analysis of continuous simulation, especially in the usual filed of application (as retention capacity lower than 10 (cm) should be avoided in practice). The possibility of considering more than one previous rainfall event into the modelling assures the quality of the results in the application of analytical probabilistic models, even in the cases of low outflow rates characterized by more than two chained rainfall events. Proposed formulas can be useful to designers, because they allow to estimate the maximum retention capacity of an IES, once the design return period has been set, and given the first moments of rainfall depth, rainfall duration and interevent time, the outflow rate, the ratio between the catchment area and the area of the infiltration system.

## REFERENCES

[1]   Marchioni, M. & Becciu, G., Experimental results on permeable pavements in urban areas: A synthetic review. *International Journal of Sustainable Development and Planning*, **10**(6), pp. 806–817, 2015. http://dx.doi.org/10.2495/SDP-V10-N6-806-817.

[2]   Pratt, C.J., Sustainable drainage: A review of published material on the performance of various SUDS components. Prepared for the Environment Agency, SUDS Science Group/99705.015, 2004.

[3]   Hogland, W., Niemczynowicz, J. & Wajlman, T., The unit superstructure during the construction period. *Science of the Total Environment,* **59**, pp. 411–424, 1987. http://dx.doi.org/10.1016/0048-9697(87)90464-5.

[4]   Pratt, C.J., Use of permeable, reservoir pavement constructions for stormwater treatment and storage for re-use. *Water Science and Technology,* **39**(5), pp. 145–151, 1999. http://dx.doi. org/10.1016/s0273-1223(99)00096-7.

[5]   Pratt, C.J., Mantle, J. & Schofield, P., UK research into the performance of permeable pavement, reservoir structures in controlling stormwater discharge quantity and quality. *Water Science and Technology*, **32**(1), pp. 63–69, 1995. http://dx.doi.org/10.1016/0273-1223(95)00539-y.

[6]   Legret, M., Colandini, V. & Le Marc, C., Effects of a porous pavement with reservoir structure on the quality of runoff water and soil. *Science of the Total Environment*, **189–190**, pp. 335–340, 1996. http://dx.doi.org/10.1016/0048-9697(96)05228-x.

[7]   Legret, M. & Colandini, V., Effects of a porous pavement with reservoir structure on runoff water: Water quality and fate of heavy metals. *Water Science and Technology*, **39**(2), pp. 111–117, 1999. http://dx.doi.org/10.1016/s0273-1223(99)00014-1.

[8]   Pagotto, C., Legret, M. & Le Cloirec, P., Comparison of the hydraulic behaviour and the quality of highway runoff water according to the type of pavement. *Water Research*, **34**(18), pp. 4446–4454, 2000. http://dx.doi.org/10.1016/s0043-1354(00)00221-9.

[9]   Asaeda, T. & Ca, V.T., Characteristics of permeable pavement during hot summer weather and impact on the thermal environment. *Building and Environment*, **35**(4), pp. 363–375, 2000. http://dx.doi.org/10.1016/s0360-1323(99)00020-7.

[10]  Schlüter, W. & Jefferies, C., Modelling the outflow from a porous pavement. *Urban Water*, **4**(3), pp. 245–253, 2002. http://dx.doi.org/10.1016/s1462-0758(01)00065-6.

[11]  Brattebo, B.O. & Booth, D.B., Long-term stormwater quantity and quality performance of permeable pavement systems. *Water Research*, **37**(18), pp. 4369–4376, 2003. https://doi.org/10.1016/S0043-1354(03)00410-X.

[12]  Dreelin, E.A., Fowler, L. & Ronald Carroll, C., A test of porous pavement effectiveness on clay soils during natural storm events. *Water Research*, **40**(4), pp. 799–805, 2006. http://dx.doi. org/10.1016/j.watres.2005.12.002.

[13] Morgenroth, J., Buchan, G. & Scharenbroch, B.C., Belowground effects of porous pavements: Soil moisture and chemical properties. *Ecological Engineering*, **51**, pp. 221–228, 2013. http://dx.doi.org/10.1016/j.ecoleng.2012.12.041.

[14] Newman, A.P., Aitken, D. & Antizar-Ladislao, B., Stormwater quality performance of a macro-pervious pavement car park installation equipped with channel drain based oil and silt retention devices. *Water Research*, **47**(20), pp. 7327–7336, 2013. http://dx.doi.org/10.1016/j.watres.2013.05.061.

[15] Marchioni, M. & Becciu, G., Infiltration-exfiltration system for stormwater runoff volume and peak attenuation. *International Journal of Safety and Security Engineering*, **8**(4), pp. 473–483, 2018. http://dx.doi. org/10.2495/SAFE-V8-N4-473-483.

[16] Teng, Z. & Sansalone J.J., In situ partial exfiltration of rainfall runoff. II: Particle separation. Journal of Environmental Engineering, 130(9), pp. 1008–1020, 2004. https://doi.org/10.1061/(ASCE)0733-9372(2004)130:9(1008).

[17] Sansalone, J.J. & Teng, Z., Transient rainfall-runoff loadings to a partial exfiltration system: Implications for urban water quantity and quality. *Journal of Environmental Engineering*, **131**(8), 2005. https://doi.org/10.1061/(ASCE)0733-9372(2005)131:8(1155).

[18] Eagleson, P.S., Climate, soil and vegetation: The distribution of annual precipitation derived from observed storm sequences. *Water Resources Research*, **14**(5), pp. 713–721, 1978. https://doi.org/10.1029/WR014i005p00722.

[19] Eagleson, P.S., Dynamics of flood frequency. *Water Resources Research*, **8**(4), pp. 878–898, 1972. https://doi.org/10.1029/WR008i004p00878.

[20] Adams, B. J., & Papa, F., *Urban Stormwater Management Planning with Analytical Probabilistic Models*, Wiley: New York, 2000.

[21] Guo, Y. & Adams, B.J., Hydrologic analysis of urban catchments with event-based probabilistic models. Part I: Runoff volume. *Water Resources Research*, **34**(12), pp. 3421–3431, 1998. https://doi.org/10.1029/98WR02449.

[22] Guo, Y. & Adams, B.J., Hydrologic analysis of urban catchments with event-based probabilistic models. Part II: Peak discharge rate. *Water Resources Research*, **34**(12), pp. 3433–3443, 1998. https://doi.org/10.1029/98WR02448.

[23] Guo, Y., Liu, S. & Baetz, B.W., Probabilistic rainfall-runoff transformation considering both infiltration and saturation excess runoff generation processes. *Water Resources Research*, **48**(6), 2012. https://doi.org/10.1029/2011WR011613.

[24] Guo, Y. & Adams, B.J., Analysis of detention ponds for storm water quality control. *Water Resources Research*, **35**(8), pp. 2447–2456, 1999. https://doi.org/10.1029/1999WR900124.

[25] Guo, Y. & Adams, B.J., An analytical probabilistic approach to sizing flood control detention facilities. Water Resources Research, 35(8), pp. 2457–2468, 1999. https://doi.org/10.1029/1999WR900125.

[26] Bacchi, B., Balistrocchi, M. & Grossi, G., Proposal of a semi-probabilistic approach for storage facility design. *Urban Water Journal*, **5**(3), pp. 195–208, 2008. https://doi.org/10.1080/15730620801980723.

[27] Balistrocchi, M., Grossi, G. & Bacchi, B., An analytical probabilistic model of the quality efficiency of a sewer tank. *Water Resources Research*, **45**(12), 2009. https://doi.org/10.1029/2009WR007822.

[28] Raimondi, A. & Becciu, G., On pre-filling probability of flood control detention facilities. *Urban Water Journal*, **12**(4), pp. 344–351, 2015. http://dx.doi.org/10.1080/1573062X.2014.901398.

[29]  Raimondi, A. & Becciu, G., On the efficiency of stormwater detention tanks in pollutant removal. *International Journal of Sustainable Development and Planning*, **12**, pp. 144–154, 2017. http://dx.doi. org/10.2495/SDP-V12-N1-144-154.

[30]  Becciu, G. & Raimondi, A., Factors affecting the prefilling probability of water storage tanks. *WIT Transactions on Ecology and the Environment*, vol. 64, WIT Press: Southampton and Boston, pp. 473–484, 2012. http://dx.doi. org/10.2495/WP120411.

[31]  Becciu, G. & Raimondi, A., Probabilistic analysis of spills from stormwater detention facilities. *WIT Transactions on the Built Environment*, vol. 139, WIT Press: Southampton and Boston, pp.159–170, 2014. http://dx.doi. org/10.2495/UW140141.

[32]  Becciu, G. & Raimondi, A., Probabilistic modeling of the efficiency of a stormwater detention facility. *International Journal of Sustainable Development and Planning*, **10**(6), pp. 795–805, 2015. http://dx.doi. org/10.2495/SDP-V0-N0-1-11.

[33]  Becciu, G. & Raimondi, A., Probabilistic analysis of the retention time in stormwater detention facilities. *Procedia Engineering*, **119**(1), pp.1299–1307, 2015. http://dx.doi.org/10.1016/j.proeng.2015.08.951.

[34]  Zhang, S. & Guo, Y., An analytical probabilistic model for evaluating the hydrologic performance of green roofs. *Journal of Hydrologic Engineering*, **18**(1), pp. 19–28, 2013. https://doi.org/10.1061/(ASCE)HE.1943-5584.0000593.

[35]  Zhang, S. & Guo, Y., Explicit equation for estimating storm-water capture efficiency of rain gardens. *Journal of Hydrologic Engineering*, **18**(12), pp.1739–1748, 2012. https://doi.org/10.1061/(ASCE)HE.1943-5584.0000734.

[36]  Guo, Y., Stochastic analysis of hydrologic operation of green roofs. *Journal of Hydrologic Engineering*, **21**(7), 2016. https://doi.org/10.1061/(ASCE)HE.1943-5584.0001371.

[37]  Raimondi, A. & Becciu, G., Performance of green roofs for rainwater control. *Water Resources Management*, 2020.

[38]  Guo, Y. & Baetz, B.W., Sizing of rainwater storage units for green building applications. *Journal of Hydrologic Engineering*, **12**(2), pp. 197–205, 2007. https://doi.org/10.1061/(ASCE)1084-0699(2007)12:2(197).

[39]  Raimondi, A. & Becciu, G., An analytical probabilistic approach to size cisterns and storage units in green buildings. *11th International Conference on Computing and Control for the Water Industry 2011 – CCWI 2011*, pp. 179–184, 2011.

[40]  Raimondi, A. & Becciu, G., Probabilistic design of multi-use rainwater tanks. *Procedia Engineering*, **70**, pp. 1391–1400. 2014. https://doi.org/10.1016/j.proeng.2014.02.154.

[41]  Raimondi, A. & Becciu, G., Probabilistic modelling of rainwater tanks. *Procedia Engineering*, **89**, pp. 1493–1499, 2014. http://dx.doi.org/10.1016/j.proeng.2014.11.437.

[42]  Becciu, G., Raimondi, A. & Dresti C., Semi-probabilistic design of rainwater tanks: A case study in Northern Italy. *Urban Water Journal*, **15**(3), pp. 192–199, 2016. http://dx.doi. org/10.1080/1573062X.2016.1148177.

[43]  Guo, Y. & Gao, T., Analytical equations for estimating the total runoff reduction efficiency of infiltration trenches. *Journal of Sustainable Water in the Built Environment*, **2**(3), 2016. https://doi.org/10.1061/JSWBAY.0000809.

[44]  Zhang, S. & Guo, Y., Stormwater capture efficiency of bioretention systems. *Water Resources Management*, **28**(1), pp. 149–168, 2014. https://doi.org/10.1007/s11269-013-0477-y.

[45]    Zhang, S. & Guo, Y., An analytical equation for evaluating the stormwater volume control performance of permeable pavement systems. *Journal of Irrigation and Drainage Engineering*, **141**(4), 2015.
http://dx.doi. org/10.1061/(ASCE)IR.1943-4774.0000810.

[46]    Howard, C.D.D., Theory of storage and treatment plant overflows. *Journal of the Environmental Engineering Division*, **102**(4), pp. 709–722, 1976.

[47]    Loganathan, G.V. & Delleur, J.W., Effects of urbanization on frequencies of overflows and pollutant loadings from storm sewer overflows: A derived distribution approach. *Water Resources Research*, **20**(7), pp. 857–865, 1984.
https://doi.org/10.1029/WR020i007p00857.

[48]    U.S. EPA, Storm water technology fact sheet: Porous pavement. Report No. EPA 832-F-99-023, Washington, DC, 1999.

[49]    Collins, K.A., Hunt, W.F. & Hathaway, J.M., Hydrologic comparison of four types of permeable pavement and standard asphalt in eastern North Carolina. *Journal of Hydrologic Engineering*, **13**(12), pp. 1146–1157, 2008.
https://doi.org/10.1061/(ASCE)1084-0699(2008)13:12(1146).

[50]    Nemirovsky, E.M., Welker, A.L. & Lee, R., Quantifying evaporation from pervious concrete systems: Methodology and hydrologic perspective. *Journal of Irrigation and Drainage Engineering*, **139**(4), pp. 271–277, 2013.
https://doi.org/10.1061/(ASCE)IR.1943-4774.0000541.

[51]    Becciu, G. & Paoletti, A., Random characteristics of runoff coefficient in urban catchments. *Water Science and Technology*, **36**(8–9), pp. 39–44, 1997.

[52]    Adams, B.J., Fraser, H.G., Howard, C.D.D. & Hanafy, M.S., Meteorological data analysis for drainage system design. *Journal of Environmental Engineering*, **112**(5), pp. 827–848, 1986.

[53]    Bedient, P.B. & Huber, W.C., *Hydrology and Floodplain Analysis*, 2nd ed., Addison Wesley: New York, 1992.

[54]    Benjamin, J.R. & Cornell, C.A., *Solutions Manual to Accompany Probability, Statistics, and Decision for Civil Engineers*, McGraw-Hill, 1970.

# USING AN ECOSYSTEM SERVICES LENS TO EXPLORE A BROADER FUNDING BASE FOR LANEWAYS

JESSICA LAMOND, MARK EVERARD & GLYN EVERETT
University of the West of England, UK

## ABSTRACT

Greening small urban street spaces such as alleys and laneways is increasingly popular with city authorities. Motivation for starting these projects varies and this may hold implications in terms of the responsible function and department overseeing planning and implementation. Knock-on effects of such decisions may also include the level and breadth of stakeholder engagement and consultation in the process, the visions created and the eventual benefits realised. This research explored the potential to widen funding sources for a laneways project in Melbourne, Australia, including crowd funding. An ecosystem services framework was used to extensively consider potential benefits from a project before and during the design phase. The findings indicate that such an approach can be very useful in order to: widen participation; tailor design to optimise benefits; bring funding from special interest groups; and increase visibility and potential for improved feedback benefits such as green tourism and property values.
*Keywords: laneways, Blue-Green Infrastructure, Green Infrastructure, ecosystems services.*

## 1 INTRODUCTION

Greening small urban pedestrian street spaces such as alleys or laneways is increasingly popular with city authorities [1], [2]. The spaces have different names but a common feature is that they provide "access to the back of properties which have a complementary street as their primary and front access" [3]. The driving motivation for starting these projects varies, however they may often include a stormwater management element within the design. Newell et al. [4] concluded that the schemes they analysed were mostly narrowly focussing on stormwater, whereas Lindt [1] recognised some schemes in the US had an economic or a social focus. Meanwhile, in Australia a process called "activation" of laneways has been part of an urban revitalisation strategy in large and small cities [2] but of the six small cities studied in Queensland, none had considered greening as a regeneration strategy. Nevertheless, the diversity urban "green infrastructure" (GI) solutions provide opportunities to optimise a wide range of ecosystem services of cumulatively significant benefit to urban residents and users and the general "liveability" of the city [5].

Multiple benefits of retrofit Blue-Green Infrastructure (BGI) such as green roofs in cities have been considered to benefit a wide range of stakeholders [6], [7] directly and indirectly beyond those generally considered and therefore consulted. However, in designing projects, a full assessment of actual or potential benefits is not generally addressed, resulting in many potential benefits and community interests being overlooked, undervalued and not realised during design and implementation [8].

Research that can inform the design of schemes towards optimising benefits is limited but indicates that perceptions and preferences of communities are important, they vary and they impact on acceptability and sustainability. While there is some research that examines the suitability of different plantings in a hydrological sense [9], very little research investigates which plants and functions communities prefer [10]. Research and best practice call for the contextualisation of benefits and the examination of specific needs and limitations locally. Consultation may be seen as onerous and a cost and delay to projects, but an alternative view is that it represents a method to bolster wider public support and also bring additional funding

WIT Transactions on The Built Environment, Vol 194, © 2020 WIT Press
www.witpress.com, ISSN 1743-3509 (on-line)
doi:10.2495/FRIAR200141

opportunities to the table by optimising interlinked benefits and tailoring design to engage and transparently benefit wider interest groups. Partnerships between local communities, interest groups and authorities naturally ensue, but defining the scope of consultees requires some prior understanding of potential beneficiaries.

This decentralisation of engagement with and funding of local natural infrastructure, comprising natural processes protected, restored or emulated to add value to engineered infrastructure is also part of a wider political discourse and policy focus. The UK government used the term "Big Government" to "Big Society" to describe the process of shifting the balance of power from central control to citizens, communities and local government, devolving powers to work together and solve localised perceived problems [11].

Historic development of Melbourne City in the Victorian era was around a grid of city streets, and between them a less predictable weave of laneways. These laneways are mostly pedestrian, many covered, and have become a significant draw for both tourists and local people due to a proliferation of multicultural food and drink, fashion and art, and other "boutique" businesses distinctive to Melbourne [12]. Between 2015–2017 the Melbourne Laneways project sought to explore the potential to "green" laneways in order to alleviate surface water and urban heat islands [13]. The research reported here was carried out over the same period to provide evidence to enable future stages of the programme. The aim of the research was to consider the range of interconnected benefits potentially arising from the greening of Melbourne's laneways in order to: map stakeholders likely to receive those benefits; to identify alternative and novel funding partners; and to support the ambition to gain some support for crowd funding future laneway greening projects.

## 2 METHOD

The approach taken is a deductive ecosystem services analysis to broaden the thinking about the stakeholders and potential funding routes for the "greening" of Melbourne's laneways. The research also drew upon already existing concepts of benefits provided by Melbourne City Council (see Table 1) [14]. The potential benefits from Laneways were then brainstormed by a team of researchers from built and natural environment, in collaboration with local experts, using the ecosystem framework to learn about potential beneficial outcomes resulting from greening and to map these to beneficiaries across a range of scales.

### 2.1 Approaching the benefits of greening from the perspectives of ecosystem services

The clustering of potential benefits under the three principal sustainable development vectors of "environmental", "economic" and "social" is helpful. However, a commonly encountered problem – frequently encountered in practical implementation of purported systems approaches – is that inherent interdependencies between these three parameters tend to be overlooked. For example, a tree planted in a laneway may confer benefits: Environmentally through providing some habitat for wildlife, cooling and cleansing the air; and slowing down flood generation; Socially, by the fact that "greener" views tend to reduce personal stress (as outlined below) as well as less polluted air having direct health benefits; and Economically, A more pleasant vista can significantly enhance local real estate values, the cleaner air can have implications for health costs, and buffering of flood risk can avert the costs of defence or damage.

It is also important to ensure that potential disbenefits are considered. For this reason, we use a framework that takes account of these systemic interdependencies as the basis for a broader assessment of linked benefits and potential disbenefits arising from different approaches to greening the laneways. As many of these potential benefits and disbenefits are vectored to and between people via ecosystem functions, we have selected the ecosystem services framework for this purpose.

Ecosystem services are defined by the UN Millennium Ecosystem Assessment (MA) as "...the benefits people obtain from ecosystems", thus they are inherently anthropocentric. However, many of the diverse services are not immediately utilitarian, some rely on value-laden, aesthetic and cultural benefits while others are related to quality of living conditions. They also encompass the integrity and resilience of ecosystems and their continued capacity to function and produce services. Specifically, we use the classification of ecosystem services developed to by the MA [15]. The MA classification of ecosystem services integrated a wide range of pre-existing ecosystem service classifications, many addressing specific habitat types and bioregional perspectives. Importantly, the MA classification explicitly spans diverse value systems across its four qualitatively different categories (see Table 2).

Table 1: Existing thinking about the Benefits of Laneway Greening published by Melbourne City [14].

The challenge of a changing climate is providing opportunities to change the way we see and create our cities. The introduction of green spaces in built up areas has many real benefits and can help cities adapt to extreme weather events, increasing urban density, population increases, reducing excess urban warming and enhancing community health and wellbeing.

Trees, vines and plants in Melbourne's laneways can provide multiple benefits such as:

**Environmental**
- diverting storm water run-off from laneways into the soil
- filtering dust and pollution from the air
- improving biodiversity levels in the central city
- the provision of habitat for wildlife
- reducing noise levels in the city
- insulating buildings from heat and cold, reducing energy expenditure and carbon emissions
- reducing the Urban Heat Island (UHI) effect through shading and cooling.

**Economic**
- insulating the building in winter and summer, reducing heating and cooling costs
- extending the life expectancy of impervious surfaces
- increasing surrounding property values
- provision of useable green outdoor space for businesses in laneways e.g. bars, cafes and restaurants.

**Social**
- provision of more green public open spaces in the central city
- reinvigorating laneways from waste areas to useable public spaces
- provision of more pleasant walkways and thoroughfares, encouraging people to walk and spend time outdoors
- reducing vandalism and antisocial behaviour
- bringing nature into the city which has positive effects in reducing depression and illness
- further promoting the City of Melbourne as a sustainable, resilient, livable city.

Table 2:  Ecosystem services category adapted from World Resources Institute [15].

- Provisioning services: food, fresh water, biochemicals and other substances and energy that can be extracted from nature;

- Regulatory services: natural processes that regulate, for example, flows and quality of air and water, erosion, diseases and climate;

- Cultural services: non-material benefits derived from nature such as spiritual enrichment, tourism and recreation opportunities, and education and research; and

- Supporting services: processes such as soil formation, photosynthesis and the cycling of nutrients and water that maintain ecosystem functioning, resilience and capacity to keep producing other more directly utilised services

All ecosystem services enhance different aspects of human wellbeing [15]. However, the supporting services are, as their name suggests, doing so indirectly through enhanced ecosystem integrity, health and by boosting the systems production of provisioning, cultural and regulatory services. Mirroring this a key element of human wellbeing, "Freedom of choice and action" depends on satisfaction of other more basic biophysical and social needs supported by ecosystem services.

This framework recognises the interconnectivity of ecosystems, the services they provide and the benefits derived. It provides a deductive tool to translate choices in technologies, and management or use of habitat, landscapes and urban spaces into changes in ecosystem vitality and functioning. Ultimately such choices can be observed impacting on an intimately connected network of social and economic benefits and disbenefits.

Several models of the ecosystem service framework have been proposed. Early precursors included habitat and location specific classifications such as Dugan [16] wetland model and the SWAMP model by Everard et al. [17], all ultimately informing the 2005 MS classification. Development of subsequent models include those of The Economics of Ecosystems and Biodiversity (TEEB) ecosystem services framework [18]; the Common International Classification of Ecosystem Services (CICES) [19] developed by the European Environment Agency (EEA); and the economic model of the UK National Ecosystem Assessment (UK NEA) [20] that seeks to monetise services as far as possible. Many of these subsequent reclassifications omit supporting services to avert double-counting of their contributions to other more directly exploited services. However, since valuation was not a feature of this study, inclusion of the fundamental roles supporting services play in productive ecosystems are vital for informing sustainable development and management strategies.

Our use of the MA classification relates to the recognition of all services, marketed and monetisable as well as non-market, as it is important to account for the full range of benefits and potential disbenefits in decision-making. Otherwise, established markets will continue substantially to distort the allocation of resources and hence the viability of productive systems. In an urban setting, financial returns from markets may today reward maximisation of built assets per unit area, yet costs in terms of heat islands, flooding, poor air quality and the psychological effects of "hard" environments may be substantial yet wholly externalised.

The analysis was then mapped against beneficiaries and stakeholders these were spatially and functionally grouped.

## 3 RESULTS

Using the ecosystem services framework as described above enabled stratification of the range of potential benefits and disbenefits from the greening for Melbourne's laneways. By taking a fully systemic approach, both recognised outcomes as well as some that may not previously have been recognised can be addressed.

### 3.1 Identifying ecosystem benefits

Potential benefits are summarised in Table 3, along with description of benefits recognised as significant.

Table 3:  Ecosystems services provided by laneways.

| Provisioning services | Description of potential provisioning service benefit(s) | Beneficiaries |
|---|---|---|
| Fresh water | Contribution to the overall water cycle, with marginal impact on water security for the city | City residents, marginally benefitting from water supply |
| Food | Benefit if planting contains some edible crops | Local people |
| Fibre and fuel | Small resource value if useful fibre plants (kapok, etc.) are included in plantings | Local beneficiaries |
| Genetic resources | A genetic resource conservation role could be served if local and scarce genetic strains form the basis for plantings | Multiple beneficiaries including future generations |
| Biochemicals, natural medicines, pharmaceuticals | A resource conservation role could be served if local and scarce strains of biochemical/medicinal value form the basis for plantings | Mainly local beneficiaries |
| Ornamental resources | Small Resource value if ornamental species are included in plantings | Local beneficiaries and visitors |
| Energy harvesting | No benefits perceived from greening | |
| Regulating services | Description of potential regulating service benefit(s) | Beneficiaries |
| Air quality regulation | Filtering dust and pollution from the air | Public health, and regulators/other with public health interests |
| Microclimate regulation | Insulating buildings from heat and cold, reducing energy expenditure and carbon emissions  Reducing the Urban Heat Island (UHI) effect through shading and cooling  Insulating the building in winter and summer, reducing heating and cooling costs | Owners and developers of buildings benefitting from improved indoor and outdoor climate regulation (health, energy costs, overcoming "sick building" syndrome, knock-on for real estate values, etc. |
| Global climate regulation | Averting the need for cooling/heating has implications for climate-active gas emissions | The international community including future generations |
| Water regulation | Diverting storm water run-off from laneways into the soil  Reduced generation of storm flooding, regulating flood risk on site and downstream  Recharging groundwater, helping buffer potential drought impacts | Beneficiaries of regulated flood and drought risk at catchment scale |

Table 3: Continued.

| Regulating services | Description of potential regulating service benefit(s) | Beneficiaries |
|---|---|---|
| Natural hazard regulation | Minor effect on slowing air speeds (which may or may not be an issue in enclosed laneways) | Local people |
| Pest regulation | Hosting of pest predators (wasps eating aphids, etc.) Hosting some pests (flies, wasps, aphids, etc.) necessitating careful management | Local people |
| Disease regulation (human) | Some benefit from air quality improvement Potentially also from attenuating airborne transmission of microbes (though this needs further testing and verification) | Local and visiting people |
| Disease regulation (stock) | None: no stock in the laneways | |
| Erosion regulation | None: no erosion in the laneways | |
| Water purification and waste treatment | Minor impact possible, but considered trivial relative to hydrological benefits | |
| Pollination | Benefit when pollinating species are hosted in "green roofs", etc. | Local people, including adjacent urban gardens |
| Salinity regulation | Not a perceived problem | |
| Fire regulation | A risk if combustible vegetation is unmanaged and allowed to dry, necessitating careful species selection and management | Local people |
| Noise and visual buffering | Reducing noise levels in the city Providing some visual screening | Local people, visitor enjoyment and real estate values |
| Cultural services | Description of potential cultural service benefit(s) | Beneficiaries |
| Cultural heritage | Inclusion of characteristic local species can imbue a local character to the laneways | Local people and visitors |
| Recreation and tourism | Can turn the laneways into an even more attractive tourism attraction and recreational space | Local people, visitors, local businesses and tourism interests |
| Aesthetic value | Provision of useable green outdoor space for businesses in laneways e.g. bars, cafes and restaurants Provision of more green public open spaces in the central city | Local people, visitors, local businesses and tourism interest |
| Spiritual and religious value | None perceived that is not addressed by "cultural" and "aesthetic" considerations above | |
| Inspiration of art, folklore, architecture, etc. | Local species and a greener environment can deepen the value of the laneways as a centre for the arts, as well as accentuating the distinctiveness of laneway architecture | Local arts and crafts interests, local businesses |

Table 3: Continued.

| Cultural services | Description of potential cultural service benefit(s) | Beneficiaries |
|---|---|---|
| Social relations | Increasing surrounding property values Reinvigorating laneways from waste areas to useable public spaces Provision of more pleasant walkways and thoroughfares, encouraging people to walk and spend time outdoors Reducing vandalism and antisocial behaviour Building local people engaging and collaborating around improving the environment of their shared spaces | Local communities, with potential direct local business and tourism "knock on" benefits |
| Educational and research | Creates "indoor green classrooms" as well as research opportunities | Local schools and universities |
| Supporting services | Description of potential supporting service benefit(s) | Beneficiaries |
| Soil formation | Benefit unlikely in this setting | |
| Primary production | Small amount of biomass production Potential disbenefit of fall of dead leaves requiring removal | Indirect benefits to biodiversity Local authorities |
| Nutrient cycling | Small/minimal benefit in this setting | |
| Water recycling | Localized evaporation and recapture by vegetation, can contribute to a cooler microclimate and other beneficial services | Local traders and tourists |
| Photosynthesis (enhanced $O_2$) | Small/minimal benefit in this setting | |
| Provision of habitat | Selection of appropriate vegetation enhances biodiversity, contributes to genetic diversity and, potentially, wider nature conservation goals | Local traders and tourists |

A significant body of ecosystem services science recognises that benefit realisation from ecosystem services can span a range of scales [21]–[23]. The section below maps some of these benefits spatially.

3.2  Mapping benefits to stakeholders

Understanding of the stakeholders benefitting from greening was derived from the ecosystem services analysis in section 3.1. These have been grouped by spatial scale and the benefit classes depicted in relation to these spatial groups on Fig. 1. It shows that many of the identified benefits are concentrated at a very local scale. This is related to the relatively small-scale adaptation of individual laneways. Larger spatial-scale benefits identified relate to genetic resources and climate regulation. These benefits may be felt at global, national and regional scale but the marginal contribution of the laneways to macro-climate cannot be seen as critical, and a contribution to genetic resources would require very careful design and planning. City-scale benefits are related to regulation of water and natural hazards. Local residents and businesses may also derive indirect benefit from these large spatial scale impacts through feedback mechanisms of tourism (if sufficiently publicised), improved property value and reduced local taxes. This analysis therefore suggests that the majority of

# Mapping of potential benefits from greening of laneways

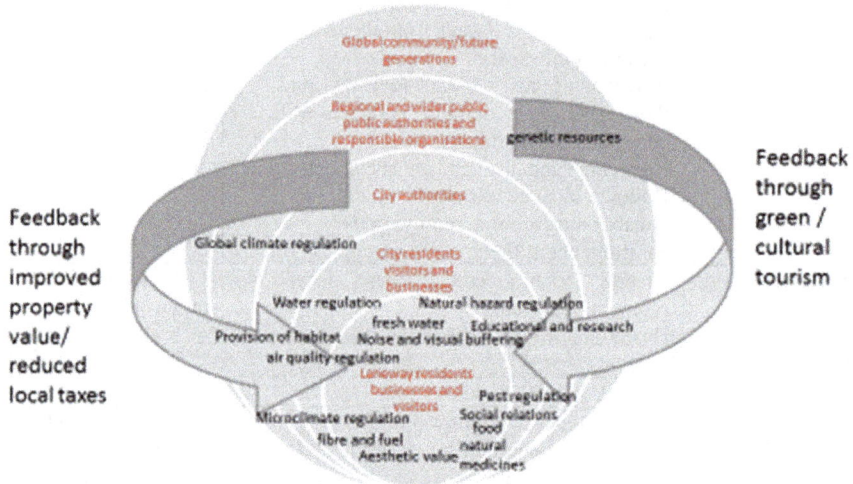

Figure 1:    Distribution of benefits of greening laneways across spatial scales including feedbacks.

the funding for laneways would sensibly be asked from local beneficiaries with some national and regional support for environmental benefits such as biodiversity protection or enhancement.

The focus on local residents and businesses as well as the potential to enhance the benefits through a focus on green and cultural tourism opens up the possibility to offer city-wide businesses and residents a chance to invest in the projects through crowd funding. Their willingness to donate will be linked to their perception of the proposed laneways, and participation in and ownership of the laneways concept and design.

## 4 DISCUSSION

In most operational contexts, a central driving disciplinary need or policy imperative tends to drive natural resource use or management decisions. The driving factor may dominate decision-making with respect to use of the resource, overlooking or dismissing other linked services. This was not the case for the laneways project, enabling a broad-ranging exploration of benefits before and during implementation, considering a broad spectrum of potential benefits and disbenefits including many commonly unaccounted or disregarded externalities. Increasing selected ecosystem services is of prime importance and generates specific societal value, but there is a risk that lack of attention to non-focal ecosystems may cause detriment that can diminish overall societal benefit. Analysis of change should be couched within an understanding of the totality of the ecosystem services provided by a habitat or ecosystem. Considering other systemic impacts, means that focal services can form an "anchor" around which other linked ecosystem service outcomes can be considered and optimised [24]. This study was limited to consideration of the introduction of new ecosystem services, other local partners in this project observing that extant ecosystems in the laneways pre-greening are minimal. Recognition that management and policy initiatives present systemic options is an important step towards increasingly systemic practice.

As both natural and managed ecosystems generate linked sets of services that cumulatively provide greater societal benefit than the sum of individual services [25], a systemic overview of outcomes from different management options is important if we are to avoid the myopia for considering single ecosystem services in isolation. These sets of linked "environmental services" have also been described as "bundles", comprising "…sets of different services that interact synergistically and occur simultaneously across landscapes provided by different land uses" [26]. Whilst not all individual aspects of ecosystem use can be fully accounted for, when considered as a system the cumulative value of multiple marketed and non-marketed services can be seen to be substantial. For example, an overall ecosystem service value for global forests has been calculated at over $16 trillion [27], of which only 6% of temperate forest and 1.6% of tropical forest valuation is from the bundled provisioning service of "raw materials" [28]. Recognition of this multiplicity of linked benefits, or the potential for the erosion of societal value where non-focal services are overlooked and commonly inadvertently degraded, is important for engaging all interests in society in collectively beneficial interventions. Systemic consideration of ecosystem services also has significant implications for social justice and net societal value [23].

In that context, systemic solutions contribute to sustainable development by allowing for a wider optimisation of benefits and the avoidance of unintended negative impacts. Ideally, this is carried out in conjunction with the evaluation of other planned or potential adjacent ecosystems service enhancements. The greening of laneways, could be considered alongside a converging range of ecosystem-based urban management technologies (including for example SuDS, or sustainable drainage systems), wetlands, and washlands, and rehabilitation of urban rivers. Further research is needed to optimise approaches across systems from the urban and into the peri-urban and to constitute systemic solutions that optimise benefits across a linked set of ecosystem services [11].

In the context of urban design, the importance of often neglected urban ecosystems and their processes are recognised as of vital importance to the viability of cities, just as resource flows supporting urban life reach out into rural and increasingly global hinterlands [20]. This point highlights the importance of recognition and internalisation of these ecosystem services into decision-making for sustainable built environments. The natural infrastructure of cities includes a complex mosaic of often overlooked ecosystem services. These include: flow paths and infiltration opportunities formed by the existing topography; drainage systems and permeability of the landscape that aid stormwater management; noise and visual mitigation through GI such as street trees and green spaces; aesthetic enhancement and amenities from BGI; and cooling through shading and evapo-transpiration to reduce urban heat. These types of beneficial habitats and processes can be emulated in microcosm through greening laneways, with a range of linked ecosystem service co-benefits. Municipal government has a lead role to play in brokering dialogue about recognition, realisation and optimisation of multiple benefits in many aspects of urban planning, design and management, the crucial factor being how it develops and appraises options in the planning process, and how it engages with multiple stakeholders comprising potential linked beneficiaries of ecosystem services.

## 5 CONCLUSION

Our analysis based on the ecosystem services framework broadened the range of benefits explored relative to traditional approaches more narrowly focused on single outcomes. In particular, it identified the potential to optimise overall benefits in the delivery of urban greening. For example, the inclusion of rare or iconic native species adapted to conditions occurring in the laneways, potentially including those with pharmaceutical properties, could create educational and other a national or regional interest in the programme, broadening its appeal and potentially attracting new sources of funding. Creative thinking about native plantings may in turn attract other native and potentially rare wildlife, creating an urban ecological haven of broad ecological, social and economic value.

We propose that by widening the consideration of benefits, using the systemic ecosystem service framework to extensively consider potential benefits and disbenefits, before and during project design phases can be very useful in order to: widen participation; tailor design to optimise benefits; provide a coherent model facilitating pooling of often fragmented outcome-specific budgets; attract novel funding from special interest groups; and increase visibility and potential for improved feedback benefits such as green tourism and property value.

All of these benefits are, in reality, intimately interconnected and must therefore be planned on an integrated basis if unintended negative consequences are to be averted and synergies are to be achieved. Mapping of the benefits on a spatial scale provides a basis for a coherent communication plan, including potential co-funding partners in the design and recognising the benefits of a collaborative approach.

## ACKNOWLEDGEMENTS
The authors acknowledge funding from NERC NE/N019180/1 and the EPSRC EP/P004237/1 as well as the collaboration of Melbourne City Council, Newcastle University, David Hetherington of ARUP and Thick Media in formative discussions.

## REFERENCES
[1]   Lindt, R., *The Avalon Green Alley Network Demonstration Project: Lessons Learned From Previous Projects for Green Alley development in Los Angeles & Beyond*, UCLA Luskin Center for Innovation, 2015.
[2]   Dillon, M., Little streets in small cities: The role of laneway activation in regional queensland cbd revitalisation. Faculty of Health, Engineering & Sciences, University of Southern Queensland, 2013.
[3]   Martin, M.D., The question of alleys, revisited. *URBAN DESIGN International*, **6**(76–92), 2001.
[4]   Newell, J.P. et al., Green alley programs: Planning for a sustainable urban infrastructure? *Cities*, **31**, pp. 144–155, 2013.
[5]   Grant, G., *Ecosystem Services Come To Town: Greening Cities by Working With Nature*, Wiley-Blackwell: Chichester, 2012.
[6]   Everett, G. & Lamond, J., Perceptions of green roofs in UK commercial real-estate. *Journal of Corporate Real Estate*, **21**(2), pp. 147–164, 2019.
[7]   Lamond, J.E., Wilkinson, S. & Rose, C., Conceptualising the benefits of green roof technology for commercial real estate owners and occupiers. *Resilient Communities, Providing for the Future, 20th Annual Pacific Rim Real Estate Conference*, PRRES: Christchurch, 2014.
[8]   Everard, M., *The Ecosystems Revolution: Co-creating a Symbiotic Future,* PIVOT Series, Palgrave, 2016.
[9]   Johnson, T., *Treenet Trials 2009: A Species Odyssey*, Mitcham, 2009.
[10]  Hoyle, H., Hitchmough, J. & Jorgensen, A., All about the 'wow factor'? The relationships between aesthetics, restorative effect and perceived biodiversity in designed urban planting. *Landscape and Urban Planning*, **164**, pp. 109–123, 2017.

[11]   Civil Exchange, *Whose Society? The Final Big Society Audit,* Civil Exchange, 2015.
[12]   Tourism Australia, Melbourne's hidden laneways, 2016. www.australia.com/en/places/vic/melbourne-hidden-laneways.html.
[13]   City of Melbourne, Greening laneways, 2020. www.melbourne.vic.gov.au/community/greening-the-city/green-infrastructure/Pages/greening-laneways.aspx. Accessed on: 9 Mar. 2020.
[14]   City of Melbourne, Benefits of green laneways, 2016. www.melbourne.vic.gov.au/community/greening-the-city/green-infrastructure/pages/greening-laneways.aspx. Accessed on: 3 May 2020.
[15]   World Resources Institute, *Ecosystems and Human Well-Being: A Framework For Assessment*, Island Press, 2003.
[16]   Dugan, P.J., *Wetland Conservation: A Review of Current Issues and Required Actions*, Gland: IUCN (World Conservation Union), 1990.
[17]   Everard, M., Denny, P. & Croucher, C., SWAMP: A knowledge-based system for the dissemination of sustainable development expertise to the developing world. *Aquatic Conservation*, **5**(4), pp. 261–275, 1995.
[18]   TEEB, The Economics of ecosystems and biodiversity: mainstreaming the economics of nature: A synthesis of the approach, conclusions and recommendations of TEEB, in the economics of ecosystems and biodiversity, 2010. http://doc.teebweb.org/wp-content/uploads/Study%20and%20Reports/Reports/Synthesis%20report/TEEB%20Synthesis%20Report%202010.pdf.
[19]   CICES, Welcome to the CICES website, 2016. http://cices.eu/.
[20]   UK NEA 2010, UK National Ecosystem Assessment. http://uknea.unep-wcmc.org/.
[21]   Fisher, B. & Turner, R.K., Ecosystem services: Classification for valuation. *Biological Conservation*, **141**, pp. 1167–1169, 2008.
[22]   Turner, R.K., Georgiou, S. & Fisher, B., *Valuing Ecosystem Services: The Case of Multifunctional Wetlands*, Earthscan Publishing: London, 2008.
[23]   Everard, M., *Ecosystems Services: Key Issues*, Routledge, 2017.
[24]   Everard, M., Nature's marketplace. *The Environmentalist*, pp. 21–23, Mar. 2014.
[25]   Schomers, S. & Matzdorf, B., Payments for ecosystem services: A review and comparison of developing and industrialized countries. *Ecosystem Services*, **6**, pp. 16–30, 2013.
[26]   Balvanera, P. et al., The links between biodiversity and ecosystem services. *Routledge Handbook of Ecosystem Services*, eds M. Potschin, R. Haines-Young, R. Fish & R.K. Turner, Routledge: London, pp. 45–61, 2016.
[27]   Costanza, R. et al., Changes in the global value of ecosystem services. *Global Environmental Change*, **26**, pp. 152–158, 2014.
[28]   de Groot, R. et al., Global estimates of the value of ecosystems and their services in monetary terms. *Ecosystem Services*, **1**, pp. 50–61, 2012.

# HYDROLOGY, ECOLOGY AND WATER CHEMISTRY OF TWO SUDS PONDS: DETAILED ANALYSIS OF ECOSYSTEM SERVICES PROVIDED BY BLUE-GREEN INFRASTRUCTURE

VLADIMIR KRIVTSOV[1,2,3], STEVE BIRKINSHAW[4], HEATHER FORBES[1], VALERIE OLIVE[5],
DAVID CHAMBERLAIN[1], JANEE LOMAX[2], JIM BUCKMAN[2], REBECCA YAHR[1],
SCOTT ARTHUR[2], KAYOKO TAKEZAWA[6] & DEREK CHRISTIE[7]
[1]Royal Botanic Garden Edinburgh, UK
[2]Heriot Watt University, UK
[3]University of Edinburgh, UK
[4]Newcastle University, UK
[5]SUERC, University of Glasgow, UK
[6]SRUC, UK
[7]Botanical Society of Scotland, UK

## ABSTRACT

Stormwater retention ponds are an important part of blue-green infrastructure, providing multiple benefits associated with flood resilience, water quality improvements, wildlife habitat creation and increases in amenity and biodiversity values. Here we compare two ponds in Edinburgh (Scotland): Oxgangs and Juniper Green. These were both established 10–15 years ago during construction of housing estates and are 3.5 km apart. The volumes of the ponds were calculated using detailed hydrographic data (obtained as part of this study). Delineation of catchments was performed using fine resolution DEM data together with details of the storm water sewer network. Hydrological and hydrodynamic modelling was carried out using the SHETRAN and CityCAT models. The presence of the ponds not only delays peak discharge after an extreme precipitation event but also reduces it rather considerably. Reductions in peak discharge and delay are much bigger for the larger Oxgangs pond, giving a 45% reduction in discharge and a 5-minute delay for a 15-minute one-in-100-year event. Data obtained on water chemistry, abundance of planktonic organisms and abundance of macroinvertebrates suggest that the increase in pollutant levels affects biological water quality and the ecosystem structure. Oxgangs pond has much higher electrical conductivity, corresponding to higher concentrations of specific elements and lower macroinvertebrate indices than Juniper Green. However, the water in Juniper Green is enriched in Ag, Pb and a number of REE, which may be related to discarded electronics. In addition to the flood resilience and water quality benefits, both ponds provide considerable amenity and biodiversity value. To date, there are 103 and 22 species of vascular plants, 20 and 16 species of bryophytes, 5 and 2 species of non-lichenised fungi and 11 and 4 species of epiphytic lichens recorded, respectively, at Oxgangs and Juniper Green. The results presented here have implications for further research and stormwater pond design and management practices.
*Keywords: SuDS, blue-green cities, SHETRAN, CityCAT, hydrological modelling, ecosystem services, water quality, biodiversity.*

## 1 INTRODUCTION

Stormwater retention ponds are an important part of blue-green infrastructure (BGI) [1], [2] and provide multiple benefits associated with flood resilience, water quality improvements, wildlife habitat creation and increases in the amenity and biodiversity values [3], [4]. Here we present case studies of two sustainable urban drainage system (SuDS) ponds located in the south-western part of Edinburgh (Scotland).

Juniper Green Pond is situated just south-west of the Edinburgh bypass in a residential area at Woodhall Millbrae (adjacent to flats 1–12), near the Water of Leith footpath, and has an area of approximately 220 m². According to Jarvie et al. [5] the pond was (re)established

WIT Transactions on The Built Environment, Vol 194, © 2020 WIT Press
www.witpress.com, ISSN 1743-3509 (on-line)
doi:10.2495/FRIAR200151

in 2005 (www.junipergreencc.org.uk/jg300-1/leaflet.html) (previously there were old mill ponds in this area when the mill was operational) and is managed by James Gibb Company.

Oxgangs Pond is located 3.5 km east of Juniper Green in a residential area adjacent to Firrhill Neuk and has a surface area of approximately 1,750 m². Jarvie et al. give the date of establishment as 2007–2010 [5]. The pond is owned by Dunedin Canmore, but management appears to be subcontracted to Water Gems (www.watergems.co.uk/), a landscaper and water features specialist based in central Scotland.

## 2 HYDROLOGY

Detailed hydrographic measurements were carried out as part of this study with 14 depth measurements at Juniper Green and 40 at Oxgangs. The Juniper Green pond has a maximum depth of 60 cm with the deeper depths towards the north and west and shallower depths towards the south and the east. Flow into the pond is from the storm water sewer network; the inlet to the pond is from the north-west corner and the outlet is in the north-east corner. The flow out from the pond is into the Water of Leith. The Oxgangs pond has a maximum depth of 100 cm with the deeper areas towards the southern and western side of the pond. Much of the rest of the pond is shallow with maximum depths of less than 25 cm. Flow into the pond is also from the storm water sewer network with two inlets into the pond, one on the southern side and one on the eastern side. The outlet is from the northern side into the Braid Burn. The estimated volumes obtained using semi-automatic interpolation of the depth measurements were 62 m³ for Juniper Green and 441 m³ for Oxgangs, respectively.

Hydrological and hydrodynamic modelling was carried out through coupling of the well-established modelling tools designed by Newcastle University [6], [7]. Firstly, delineation of the catchments was carried out using fine-resolution DEM data together with details of the storm water sewer network (provided by Scottish Water). At Juniper Green the 2 m resolution Scottish Public Sector LiDAR (Phase I) dataset was used. At Oxgangs a detailed LiDAR survey was carried out before the estate was built, however, this was unusable as all the elevations changed considerably during the construction of the estate. The best available dataset was the 5 m Ordnance Survey DEM. Secondly, SHETRAN hydrological simulations were carried out for 19 months from 1 January 2018 to 31 July 2019 using daily SEPA rainfall measured at Torduff and the Royal Botanic Garden Edinburgh. These dates correspond to the period of the observations as part of the ecological studies. Average monthly potential evaporation was used, with the data obtained from the CHESS dataset. Also included in the model were appropriate properties of the catchments (soils, vegetation, impervious areas). The application of SHETRAN produced time series for free surface evaporation, evapotranspiration, soil moisture content, surface and subsurface runoff and pond discharges. Thirdly, hydrodynamic modelling of extreme events by the CityCAT hydrodynamic model was carried out for both catchments using two sets of contrasting input data, corresponding to the SHETRAN outputs for periods following relatively dry spells in the summer (with most of the vegetated catchment areas being unsaturated) and periods with high precipitation in winter (with most of the vegetated catchment areas being at field capacity).

The time series of water discharges (simulated using SHETRAN) follow the same pattern for both ponds. Generally, the absolute discharge (expressed in m³/s) is higher in Oxgangs, reflecting the bigger catchment. However, when expressed in mm/day (normal units to compare catchments of different areas), the discharge from the Juniper Green pond is mostly higher (Fig. 1). There are more peaks and it reflects smaller evapotranspiration figures (due to the smaller area of the open water and because of the high proportion of its catchment that is impervious). However, there are a number of exceptions due to the different rainfall in the two catchments. The nominal residence time for each month is calculated as the pond volume divided by discharge. For Oxgangs the average residence time is 10.7 days, whereas for Juniper Green it is 5.4 days. The larger value at Oxgangs reflects the larger pond compared to the size of the contributing catchment area.

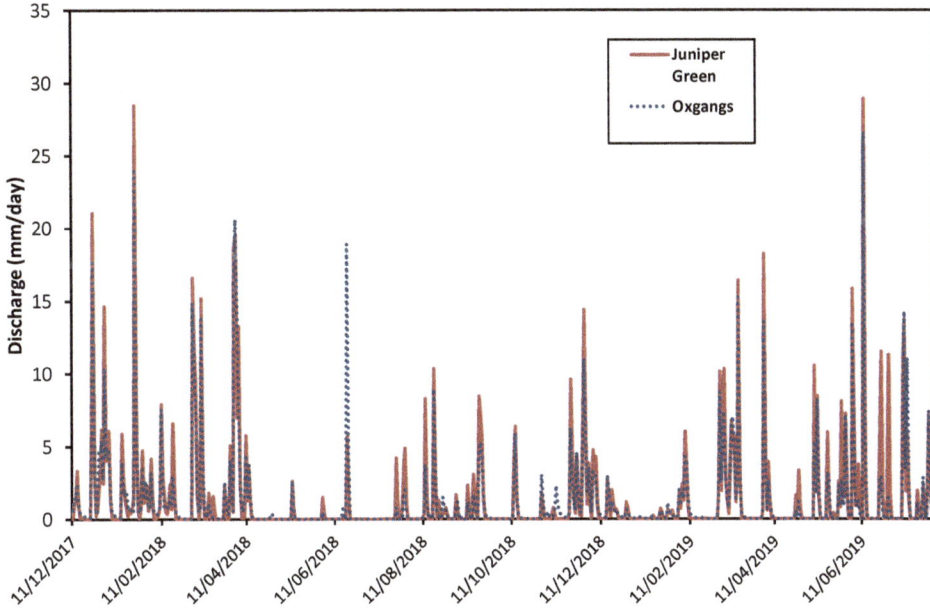

Figure 1:  Simulated discharges from the two SuDS ponds.

Higher discharge values in winter result from low evapotranspiration, whilst lower discharge values in summer and early autumn correspond to higher evapotranspiration. Consequently, the smaller through-flows in summer lead to the larger water residence times (from June to August in 2018 the average residence times at Oxgangs and Juniper Green are 41.9 and 17.1 days, respectively). This in turn decreases the rates of algal washout losses, increases sedimentation of the suspended particles and influences a range of ecosystem processes including pollution transport and biogeochemical cycling.

Table 1:  Oxgangs peak discharges and delayed time to peak (compared with no pond) for different storm durations for a one-in-100-year event with dry initial conditions.

| Storm duration | Total rainfall (mm) | Peak rainfall rate (mm/hr) | No pond | With pond | |
|---|---|---|---|---|---|
| | | | Peak discharge ($m^3$/s) | Peak discharge ($m^3$/s) | Delay (minutes) |
| 15 minutes | 24 | 227 | 0.53 | 0.29 | 5 |
| 30 minutes | 31 | 182 | 0.65 | 0.39 | 4 |
| 1 hour | 38 | 136 | 0.63 | 0.42 | 4 |
| 2 hour | 46 | 94 | 0.52 | 0.39 | 4 |
| 3 hour | 52 | 74 | 0.44 | 0.36 | 5 |
| 6 hour | 65 | 47 | 0.30 | 0.26 | 5 |

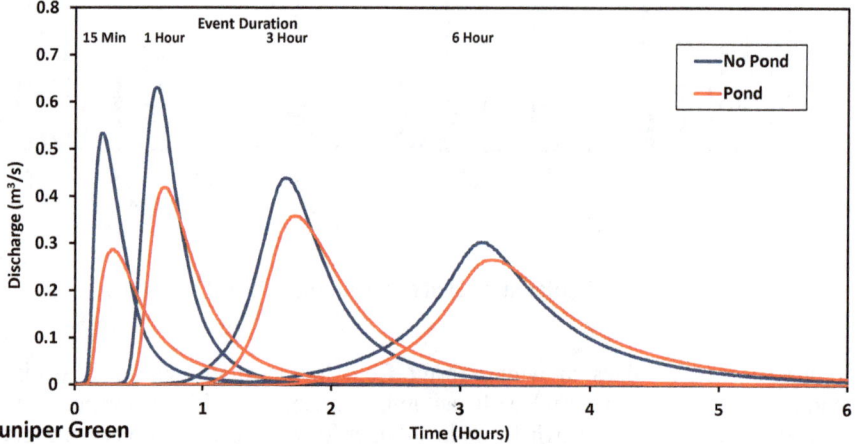

Table 2:   Juniper Green peak discharges and delayed time to peak (compared with no pond) for different storm durations for a one-in-100-year event with dry initial conditions.

| Storm duration | Total rainfall (mm) | Peak rainfall rate (mm/hr) | No pond Peak discharge (m³/s) | With pond | |
|---|---|---|---|---|---|
| | | | | Peak discharge (m³/s) | Delay (minutes) |
| 15 minutes | 24 | 227 | 0.16 | 0.14 | 2 |
| 30 minutes | 31 | 182 | 0.18 | 0.16 | 1 |
| 1 hour | 38 | 136 | 0.16 | 0.15 | 1 |
| 2 hour | 46 | 94 | 0.13 | 0.12 | 1 |
| 3 hour | 52 | 74 | 0.11 | 0.10 | 1 |
| 6 hour | 65 | 47 | 0.07 | 0.07 | 1 |

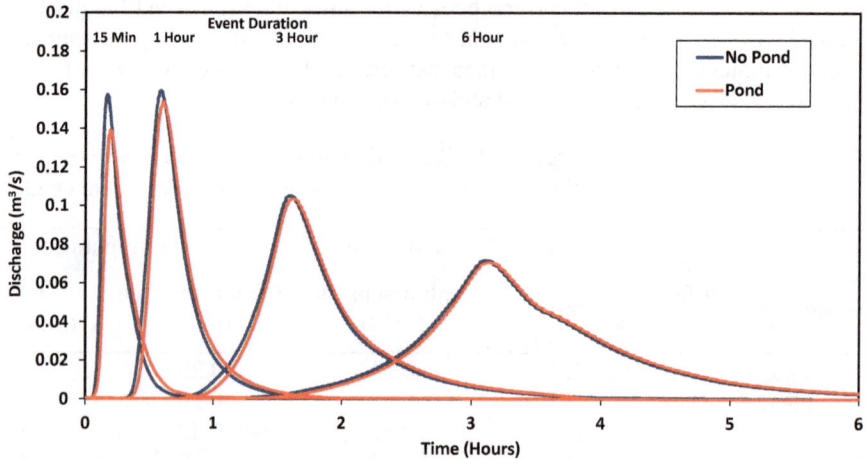

Figure 2:   Simulated CityCAT discharges at (a) Oxgangs; and (b) Juniper Green, for one-in-100-year events of four different durations with dry initial conditions.

The presence of the ponds not only delays peak discharge after an extreme precipitation event, but also reduces it rather considerably. Tables 1 and 2 and Fig. 2 show the effect of ponds at Juniper Green and Oxgangs for one-in-100-year events of different durations with dry initial conditions. The largest percentage reduction in peak discharges in both ponds was for the 15-minute event; this was a 12% reduction at Juniper Green and 45% at Oxgangs. The highest actual difference in flows was 0.26 m$^3$/s at Oxgangs for the 30-minute event and 0.018 m$^3$/s at Juniper Green for the 15-minute event. Both ponds caused smaller changes in peak discharges for the longer, less intense events. The reduction in peak discharges and the delay to the peak discharge are much larger in Oxgangs compared to Juniper Green. The main reasons for this are considered in the discussion and concluding remarks.

As expected, wet initial conditions in the CityCAT simulations increase the peak discharges. These differences are much more prominent for Oxgangs, where the percent of previous green areas constitutes a larger proportion of the catchment and it increases the peak discharge for a 15-minute event by up to 25%. For Juniper Green (with a predominantly impervious catchment) the differences in discharges between the two sets of conditions are much smaller with an increase in peak discharge for a 15-minute event of up to 5%.

A comparison of the CityCAT simulated maximum water depths at Juniper Green for the 15-minute one-in-100-year event are shown in Fig. 3. In Fig. 3(a) with no pond the water depths are generally quite small, although there is a build-up of water behind some of the buildings, whereas in Fig. 3(b) there are also large water depths corresponding to the presence of the pond. As expected, away from the pond the water depths are the same in both simulations. The CityCAT simulations for Oxgangs show a similar response with the presence of a pond just changing the depths in that part of the catchment.

## 3 WATER QUALITY AND HYDROBIOLOGY

Water chemistry (assessed using field sensors and ICP MS analysis of water samples), the abundance of planktonic organisms (sampled using a plankton net) and the abundance of macroinvertebrates (3 minutes sweep sample) were monitored between April 2018 and May 2019. Generally, the pond water in Oxgangs is characterised by a higher amount of dissolved substances and has higher electrical conductivity compared to Juniper Green. This corresponded to a number of elemental concentrations being significantly higher in Oxgangs water samples, including B, Ba, Ca, Mg, Na, Se, Sr, U and Eu. In addition, a number of further elements had dissolved concentration levels considerably higher (albeit non-significantly) in Oxgangs, including K, Li, P, Rb, Sb, Se, Si and Lu. Slightly higher values were also noted for Tm. However, this pattern was reversed for Ag, Pb and Fe, with the former two elements having significantly higher concentrations at Juniper Green (the differences in Fe concentrations, although appearing very substantial, were not significant due to an overlap in ranges). The levels of Zn, La, Ce, Pr and Nd were also considerably higher at Juniper Green, although the differences were not significant. Slightly higher values in Juniper Green were also noted for Tb and Dy. The overall higher amounts of dissolved substances at Oxgangs may be explained by its bigger and more diverse catchment. Both sites appear to be experiencing the impact of polluted runoff, but overall it is greater at Oxgangs. The enrichment of Juniper Green water in certain substances (including Ag, Pb and a number of REE) may be related to discarded electronics.

### 3.1 Macroinvertebrates

The macroinvertebrates in both ponds are mainly represented by animals tolerant of a wide range of environmental conditions (e.g. *Asellus aquaticus, Radix baltica,* Chironomidae,

a) No Pond

b) With Pond

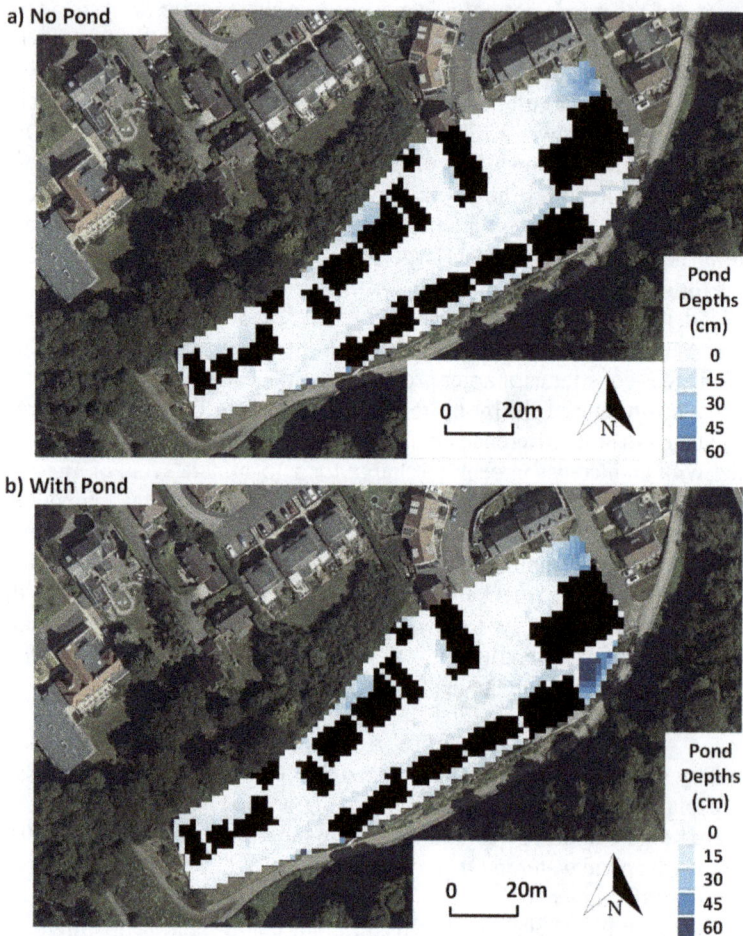

Figure 3:   Simulated maximum water depths at Juniper Green for a 15-minute one-in-100-year event with dry initial conditions. (a) With no pond; and (b) With a pond. The buildings are shown in black.

Corixidae, Planorbidae and Coenagrionidae). However, both ponds have Limnephilidae, and the Juniper Green pond also has Phryganeidae (these families are indicative of medium quality conditions). It should also be noted that the trophic level structure differs between the ponds. The predatory larvae of *Chaoborus* have not been observed in Oxgangs but are regularly encountered in Juniper Green, sometimes in rather large quantities. Also, the predatory hemipteran *Notonecta glauca* is common in Juniper Green but is rather scarce in Oxgangs. The scarcity of insect predators in Oxgangs is likely to be related to the presence of the fish *Gasterosteus aculeatus,* which is absent in Juniper Green. However, the Juniper Green pond features a healthy population of palmate newts *Lisotriton helvetica*.

From the examination of boxplots for ASPT and WHPT indices (Fig. 4), there appears to be an indication that biological water quality in Juniper Green is somewhat better than in Oxgangs, which tallies well with the water chemistry data. However, these differences are not statistically significant using the Kruskal–Wallis test.

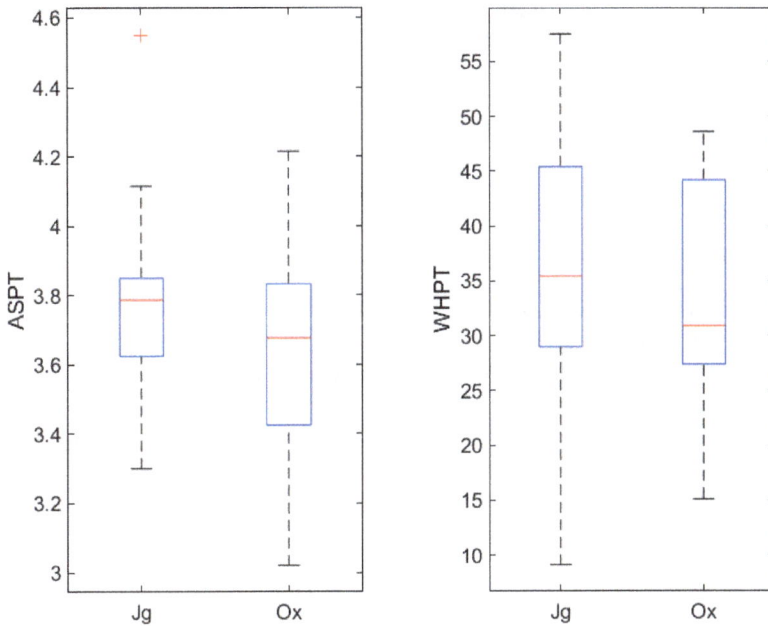

Figure 4:   Boxplot comparisons of macroinvertebrate water quality indices in Juniper Green (Jg) and Oxgangs (Ox) stormwater retention ponds.

## 3.2 Plankton

For most of the sampling period, the phytoplankton community in Juniper Green pond was dominated by *Spirogyra*, with the dinoflagellate *Peridinium* being subdominant. However, their abundance was low at the very beginning of sampling in May 2018 and also plummeted sharply at the end of winter. *Spirogyra* was not encountered in February 2019 samples and was rare thereafter. *Peridinium* was rare in February and March, but recovered by mid Spring, and was frequent in April and May 2019 samples. The demise of planktonic populations at the end of winter is likely to be a regular feature of this small pond.

Cyanobacteria (mainly *Microcystis*) were occasional in the Juniper Green samples from June, September and October, whilst diatoms were also occasionally encountered and were particularly diverse in June. The zooplankton community was dominated by different species on different sampling occasions e.g. *Daphnia* was frequent in July and October, copepods in August 2018 and May 2019, whilst protozoa were occasional throughout the summer period. The most stable occurrence, however, was observed for the rotifer *Keratella quadrata*, which was frequent or abundant for most of the investigation period. A number of other rotifers, including e.g. *Polyarthra dolichoptera*, were found in the August sample.

The yearly dynamics of the planktonic community in Oxgangs pond appeared to have a minimum in early Summer. In June, the phytoplankton community was rather sparse and represented by diatoms. Their abundance increased in further months, with *Cocconeis*, *Epithemia* and *Synedra* frequently encountered in the samples. The largest diatom diversity was revealed in the November and April 2019 samples. These peaks appear to correspond to the spring and autumn diatom blooms well-known from other temperate lentic water bodies [8].

Cyanobacteria in Oxgangs were present from July onwards but never dominated the community. However, *Oscillatoria* was frequent in both July and November samples. Green filaments were prominent from July onwards, with *Tribonema* being abundant in September and November, *Mougeotia* from September to November, and Spirogyra in July. *Mougeotia* and *Spirogyra* were also frequent at the end of sampling sequence in May 2019. Protozoa were always present; in particular *Centropyxis* was frequent in September and October samples. Rotifers were present from August onwards but have never been observed in large numbers. Cyclops were present in large numbers in the majority of the Oxgangs samples, whilst *Daphnia* were frequent in August and *Chydorus* in August, November and May 2019.

## 4  FURTHER AMENITY AND BIODIVERSITY BENEFITS

It should be noted that in addition to the flood resilience and water quality benefits, both ponds provide some considerable amenity and biodiversity value. Waterfowl have been frequently observed at Oxgangs, whilst Juniper Green benefits from the established newt *Lisotriton helvetica* population. To date, there are 103 and 22 species of vascular plants, 20 and 16 species of bryophytes, five and two species of non-lichenised fungi and 11 and four species of epiphytic lichens recorded at Oxgangs and Juniper Green, respectively. The next section gives further details of the sites' vegetation and their mycological communities.

### 4.1  Vascular plants

103 vascular plant species were recorded from Oxgangs pond. These consist of a mixture of native and non-native trees and shrubs, aquatic or mesic species and a large number of terrestrial herbaceous plants. As this pond was constructed as a SuDS feature for the surrounding housing estate and is privately managed, many of these species are likely to have been planted. The Braid Burn does, however, run close by so plants may be able to spread into the pond area from there.

The trees and shrubs comprised the native species *Betula pendula, Hedera helix, Ilex aquifolium, Prunus avium, Rosa* sp., *Rubus fruticosus* agg., *Rubus idaeus, Salix* sp. and *Sorbus aucuparia* and the non-natives *Acer pseudoplatanus, Berberis thunbergii, Buddleja davidii, Escallonia* sp., *Pyracantha* sp., *Rosa rugosa, Symphoricarpos albus* and *Weigela florida.*

Of the herbaceous species, aquatic or mesic plants included *Ceratophyllum demersum, Elodea nuttalli, Iris pseudoacorus, Lagarosiphon major, Lemna minor, Lycopus europaeus, Mentha aquatica, Menyanthes trifoliata, Potamogeton natans, Ranunculus flammula, Ranunculus lingua, Rorippa* sp. and *Typha latifolia.* Grasses were well-represented by *Arrhenatherum elatius, Agrostis capillaris, Bromus* sp., *Dactylis glomerata, Elymus repens, Festuca rubra, Glyceria maxima, Holcus lanatus, Lolium perenne, Phalaris arundinacea* and *Poa annua.* Pteridophytes were less well-represented with *Dryopteris* sp., *Equisetum arvense* and *Polypodium* sp. Of the remaining species, many were common natives that may be spontaneously-occurring or perhaps (for some) sown: *Anthriscus sylvestris, Atriplex patula, Bellis perennis, Capsella bursa-pastoris, Cardamine hirsuta, Centaurea nigra, Cerastium fontanum, Cirsium arvense, Cirsium vulgare, Digitalis purpurea, Epilobium hirsutum, Epilobium* sp., *Galium aparine, Geranium robertianum, Geum urbanum, Heracleum sphondylium, Jacobaea vulgaris, Lapsana communis, Lotus pedunculatus, Myosotis arvensis, Plantago lanceolata, Plantago major, Polygonum aviculare, Ranunculus repens, Rumex obtusifolius, Sagina apetala, Sagina procumbens, Senecio vulgaris, Sinapis arvensis, Sisymbrium officinale, Sonchus asper, Sonchus oleraceus, Stellaria graminea, Stellaria media, Taraxacum* agg., *Trifolium pratense, Trifolium repens, Tripleurospermum maritimum*, *Tussilago farfara*, *Urtica dioica* and *Vicia hirsuta*. Other, non-native, species are more probably naturalised e.g. *Aster* sp., *Calendula officinalis*, *Erysimum* sp., *Foeniculum vulgare*, *Matricaria discoides* and *Mimulus* sp. Several *Sedum* species are present around the border of the pond. The location of *Vinca major* in the ornamental shrub bed suggests planted origin. *Narcissus pseudonarcissus* is found close to the border of the pond.

Fewer vascular plant species were recorded from the site at Juniper Green, likely because of the smaller size of the site. A number of these were aquatic and mesic, such as *Alisma plantago-aquatica*, *Callitriche stagnalis*, *Caltha palustris*, *Carex pendula*, *Carex pseudocyperus*, *Crassula* sp., *Iris pseudoacorus*, *Juncus articulatus*, *Juncus effusus*, *Nymphaea alba*, *Phragmites australis* and *Ranunculus lingua*, most of these being native. Since the pond was constructed/reconstructed along with the housing estate, many of these species are likely to have been planted, although the Water of Leith is close to the pond so could be a possible source of propagules.

The other species at the Juniper Green site are mostly non-native shrubs and trees (*Cornus* sp., *Cotoneaster horizontalis*, *Cotoneaster salicifolia*, *Cotoneaster* sp., *Picea* sp. and *Rosa* sp.), with *Equisetum arvense*, *Festuca rubra* and *Hedera helix* also present.

## 4.2 Bryophytes

Twenty bryophyte species have been recorded from the Oxgangs pond site, predominantly mosses.

A range of mosses was recorded in a variety of microhabitats: *Barbula convoluta* var. *convoluta*, *Brachythecium albicans*, *Brachythecium rutabulum*, *Bryoerythrophyllum recurvirostrum*, *Bryum argenteum*, *Bryum capillare*, *Bryum dichotomum*, *Calliergonella cuspidata*, *Didymodon insulanus*, *Didymodon rigidulus*, *Hypnum cupressiforme*, *Kindbergia praelonga*, *Phascum cuspidatum*, *Polytrichum juniperinum*, *Pseudocrossidium hornschuchianum*, *Rhytidiadelphus squarrosus*, *Sanionia uncinata* and *Tortula muralis*. The liverwort *Marchantia polymorpha* subsp. *ruderalis* was also present.

Fifteen bryophyte species were recorded from Juniper Green, again mostly mosses but with three liverwort species and the unusual find of hornwort *Phaeoceros laevis*. It should be noted that *P. laevis* is rarely recorded from Scotland. The fluctuating water level that is a feature of these SuDS ponds maintains the open, moist mud margins that are shaded below the building walls. These environmental factors mirror those of the only other known Lothians site, adjacent to Inverleith House in the Royal Botanic Garden Edinburgh. Hence despite its small size, this SuDS asset provides a very important contribution to the local biodiversity, which is in line with other studies on the Edinburgh BGI ponds [9].

The mosses *Brachythecium rutabulum*, *Bryum capillare*, *Didymodon insulanus*, *Didymodon rigidulus*, *Grimmia pulvinata*, *Orthotrichum anomalum*, *Schistidium crassipilum* and *Syntrichia ruralis* subsp. *ruralis* were found on walls, while *Calliergonella cuspidata* and *Pohlia wahlenbergii* var. *wahlenbergii* were found on muddy ground. *Kindbergia praelonga* was present on bare soil.

The leafy liverwort *Lophocolea bidentata* was also present on bare soil, as was the thallose liverwort *Marchantia polymorpha* subsp. *ruderalis* and *Phaeoceros laevis*.

## 4.3 Fungi

Fungal records were collected ad-hoc during vegetation surveys so only a small number of species were recorded for both ponds. Further surveying will no doubt reveal more species present at both sites.

The basidiomycetes found around Oxgangs pond were *Panaeolina foenisecii* and *Entoloma* sp., with rust fungi represented by *Coleosporium tussilaginis* (host *Tussilago farfara*) and *Puccinia lagenophorae* (host *Senecio vulgaris*). The ascomycete *Rhytisma acerinum* was also present.

Only two basidiomycetes – *Omphalina* sp. and c.f. *Arrhenia spathulata* (NB the latter record requires checking/confirmation) – were noted at Juniper Green. The relative scarcity of records at this site reflects its small size and the availability of substrates.

## 4.4  Lichens

A survey of epiphytic lichens at Oxgangs found *Halecania viridescens, Lecanora chlarotera, Lecanora compallens, Lecanora* sp., *Lecidella elaeochroma, Physcia adscendens, P. aipolia, P.* sp., *P. tenella, Xanthoria parietina* and *X. polycarpa* on mix of native and non-native tree and shrub hosts.

The epiphytes *Lecanora* sp., *Physcia* sp., *Porina aenea* and *Xanthoria parietina* were recorded at Juniper Green, again on a mix of native and non-native tree and shrub hosts.

## 5  DISCUSSION AND CONCLUDING REMARKS

The research presented here highlights the importance of ecosystem services provided by the SuDS ponds studied and gives an account of their hydrology, ecology and water quality, as well as an insight into further multiple benefits associated with amenity and biodiversity values. The results are in line with other studies demonstrating that BGI pond sites provide an increase in flood resilience and improvements in water quality as well as aid to the creation of wildlife corridors, thus contributing to the enhancement of urban biodiversity [9].

It should be noted that the multiple benefits provided by the ponds are interconnected [3], [10] and the overall functioning of these engineered assets is best understood by considering the separate ecosystem components in concert. For instance, water quality improvements provided by the ponds are intrinsically dependent on the sites' hydrology (including e.g. precipitation patterns, through-flow and retention times), hydrography and catchment characteristics, as well as their biological community. The latter, in turn, is influenced by the hydrology and water chemistry (see, for example, the examples related to the hornwort and macroinvertebrates described above). Furthermore, there are many more aspects and indirect interactions [11], [12] beyond those specifically addressed by the present publication. For instance, biodiversity of the sites depends on the surrounding area and most importantly on their vegetation e.g., the presence of *Rhytisma acerinum* on *Acer pseudoplatanus* leaves at Oxgangs is likely on substrates blown in from adjacent areas (as well as supplied by a couple of small saplings present on site). The leaves also end up in the water, thus bringing in allochthonous detritus and associated pollutants from intercepted airborne particulates.

Comparing the two ponds shows that Juniper Green has slightly better water quality than Oxgangs, which may be explained by its smaller catchment and retention time. Also, despite its very small size this site provides a very significant contribution to the local biodiversity, featuring the presence of a rare bryophyte among a range of other recorded taxa. The biodiversity contribution from the Oxgangs site is also considerable. Furthermore, both ponds are effective at improving the flood resilience (peak discharges and delay in the peak). However, the pond at Oxgangs is much more effective at reducing the peak discharge compared to the one at Juniper Green (for the 15-minute event there was a 45% reduction in peak discharge compared to 12% at Juniper Green). As the ponds are full at the start of the simulation event this improvement is achieved by reducing the velocities. This improvement at Oxgangs is partly related to the larger pond volume and area compared to Juniper Green and also to the longer residence time at Oxgangs (10.7 and 5.4 days, respectively). However, this does not completely account for the peak discharge reduction. It appears the shape of the pond, its bathymetry and the location of the inlets and outlets have a significant effect on the how well the pond increases flood resilience. It is suggested that further work is carried out considering how the pond design affects the flood resilience.

The research presented here provides an important contribution to the case studies of hydrology, biodiversity and ecosystem services provided by SuDS and highlights the importance of comprehensive consideration of their subsystems. It should, therefore, be of use for further investigations as well as development of BGI management practices. The study is also relevant for understanding short-term and long-term environmental effects [13] and may be of value for improving the public perception of these valuable engineered assets [14].

## ACKNOWLEDGEMENTS
This study was supported by the EPSRC funding for the "Urban Flood Resilience" project (grants EP/P004180/1 and EP/P003982/1). Alejandro Sevilla, Alice Masip, Achiraya Kraiphet, Yamina Monteiro, Simon Kennedy, Cameron Diekonigin, Caroline Cruickshanks and Cesare Pertusi are kindly thanked for their various contributions to fieldwork, data processing and identification/biological recording. Help of Garth Foster was invaluable in identifying water beetles.

## REFERENCES
[1]  Brears, R.C., *Blue and Green Cities: The Role of Blue-Green Infrastructure in Managing Urban Water Resources*, Springer, 2018.
[2]  DeBarry, P.A., Addressing Italy's urban flooding problems through the holistic watershed approach by using blue/green infrastructure. *UPLanD-Journal of Urban Planning, Landscape and Environmental Design*, 4(1), pp. 127–136, 2019.
[3]  Krivtsov, V., Arthur, S., Allen, D. & O'Donnell, E., *Blue-green Infrastructure: Perspectives on Planning, Evaluation and Collaboration*, CIRIA: London, 2019.
[4]  Krivtsov, V. et al., *Monitoring and Modelling SUDS Retention Ponds: Case Studies from Scotland*, ICONHIC: Chania, Greece, 2019.
[5]  Jarvie, J., Arthur, S. & Beevers, L.J.W., Valuing multiple benefits, and the public perception of SuDS ponds, 9(2), p. 128, 2017.
[6]  Ewen, J., Parkin, G. & O'Connell, P.E., SHETRAN: Distributed river basin flow and transport modeling system. *Journal of Hydrologic Engineering*, 5(3), pp. 250–258, 2000.
[7]  Glenis, V., Kutija, V. & Kilsby, C.G., A fully hydrodynamic urban flood modelling system representing buildings, green space and interventions. *Environmental Modelling Software*, 109, pp. 272–292, 2018.
[8]  Goldman, C.R. & Horne, A.J., *Limnology*, McGraw-Hill, 1983.
[9]  Krivtsov, V. et al., Flood resilience, amenity and biodiversity benefits of an historic urban pond. *Philosophical Transactions of the Royal Society A*, 378(2168), 20190389, 2020.
[10] O'Donnell, E. et al., The blue-green path to urban flood resilience. *Blue-Green Systems*, 2(1), pp. 28–45, 2020.
[11] Krivtsov V. Investigations of indirect relationships in ecology and environmental sciences: a review and the implications for comparative theoretical ecosystem analysis. *Ecological Modelling*, 174(1–2), pp. 37–54, 2004.

[12]  Krivtsov, V., Indirect effects in ecology. *Encyclopedia of Ecology*, eds S.E. Jorgensen & B.D. Fath, Newnes, pp. 1948–1958, 2008.
[13]  Ahilan, S., et al., Modelling the long-term suspended sedimentological effects on stormwater pond performance in an urban catchment. *Journal of Hydrology*, **571**, pp. 805–818, 2019.
[14]  Williams, J., Jose, R., Moobela, C., Hutchinson, D., Wise, R. & Gaterell, M., Residents' perceptions of sustainable drainage systems as highly functional blue green infrastructure. *Landscape Urban Planning*, **190**, 103610, 2019.

# Author index

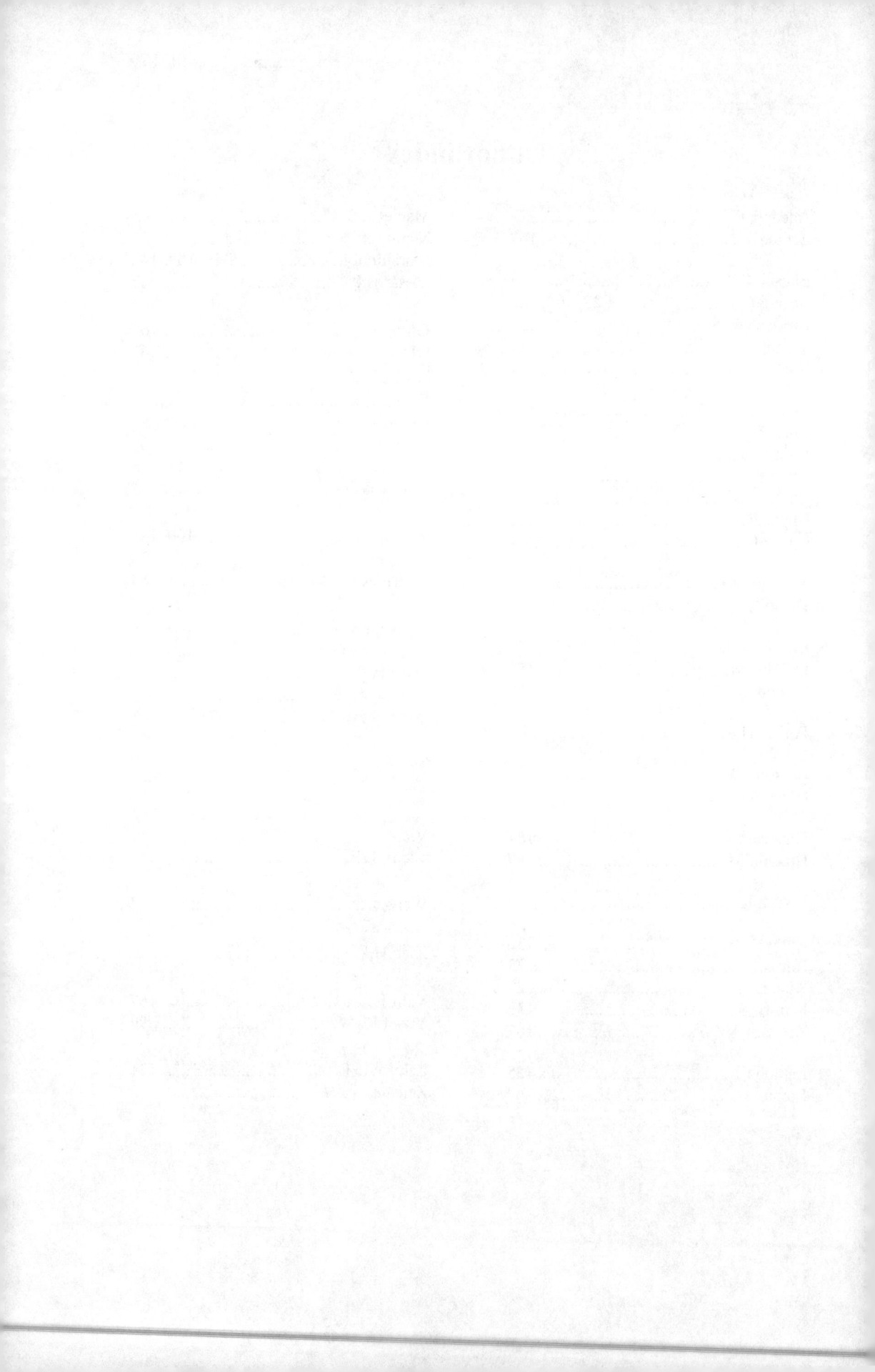

**WIT**PRESS  ...for scientists by scientists

## Water Pollution XV

*Edited by:* **S. MAMBRETTI**, *Polytechnic of Milan, Italy and* **J. S. PALENCIA JIMENEZ**, *Polytechnic University of Valencia, Spain*

Environmental problems caused by the increase of pollutant loads discharged into natural water bodies requires the formation of a framework for regulation and control. This framework needs to be based on scientific results that relate pollutant discharge with changes in water quality. The results of these studies allow the industry to apply more efficient methods of controlling and treating waste loads, and water authorities to enforce appropriate regulations regarding this matter.

Water pollution problems are essentially interdisciplinary. Engineers and scientists working in this field must be familiar with a wide range of issues including the physical processes of mixing and dilution, chemical and biological processes, mathematical modelling, data acquisition and measurement, to name but a few. In view of the scarcity of available data, it is important that experiences are shared on an international basis. Thus, a continuous exchange of information between scientists from different countries is essential.

Papers presented at Water Pollution 2020, the 15th International Conference in the series of Monitoring, Modelling and Management of Water Pollution, are contained in this volume and highlight research works from scientists, managers and academics from different areas of water contamination..

ISBN: 978-1-78466-383-4    eISBN: 978-1-78466-384-1
Published 2020 / 190pp

www.ingramcontent.com/pod-product-compliance
Lightning Source LLC
Chambersburg PA
CBHW062005190326
41458CB00009B/2976